Cave Art and Climate Change

Also by this author:

Earth Resources and Environmental Impacts

Cave Art and Climate Change

KIERAN D. O'HARA

Copyright © 2014 Kieran D. O'Hara.

All rights reserved. No part of this book may be used or reproduced by any means, graphic, electronic, or mechanical, including photocopying, recording, taping or by any information storage retrieval system without the written permission of the publisher except in the case of brief quotations embodied in critical articles and reviews.

Archway Publishing books may be ordered through booksellers or by contacting:

Archway Publishing
1663 Liberty Drive
Bloomington, IN 47403
www.archwaypublishing.com
1-(888)-242-5904

Because of the dynamic nature of the Internet, any web addresses or links contained in this book may have changed since publication and may no longer be valid. The views expressed in this work are solely those of the author and do not necessarily reflect the views of the publisher, and the publisher hereby disclaims any responsibility for them.

Cover: Bison from Altamira ceiling, Spain, and GISP2 methane climate curve.
Sources: Thinkstock.com and NOAA.

Any people depicted in stock imagery provided by Thinkstock are models, and such images are being used for illustrative purposes only. Certain stock imagery © Thinkstock.

ISBN: 978-1-4808-1130-0 (sc)
ISBN: 978-1-4808-1131-7 (e)

Library of Congress Control Number: 2014916531

Printed in the United States of America.

Archway Publishing rev. date: 10/3/2014

Contents

Preface vii

1. Introduction 1
2. Who Were the Cave Painters? 9
3. A Journey Through the Ice Age 23
4. Ice-Age Herbivores 37
5. Cold- and Warm-Adapted Animal Groups 49
6. Techniques and Dating of Cave Art 63
7. The Cave Art Fallacy 77
8. Cave Art and the Climate Curve 87
9. Mammoth Migrations 99
10. Cave Imagery as Cultural Meme 105

Index 117

Preface

This book is written with a general audience in mind, an audience with no background in Upper Paleolithic European (French-Spanish) cave art. You may be glad to know that the text contains only one equation (in chapter 8), but it does contain numerous graphs, and therefore it assumes some scientific literacy. It is also not a coffee-table book, the only color image being on the cover. *Cave Art* by Jean Clottes is a well-illustrated reasonably priced book and would be a useful companion to this book for those not familiar with cave imagery.

I use the terms cave art and cave imagery interchangeably throughout the text. Apart from the "six giants" illustrated in Henri Breuil's book *Four Hundred Centuries of Cave Art* (1952), most cave representations or depictions are best described simply as images; most of them are drawn accurately enough so that the animal depicted can be identified, but they might not qualify as art as we currently understand that term. If a highly refined technique is a requirement, then the six giants certainly qualify as art, and the Chauvet and Cosquer caves could also be added. An alternative definition of art is that the image was conscious of itself—namely the painter conceived that others might also look at it. This clearly is the case in numerous examples where the topography or curvature of the cave wall rock was used to enhance the drawing, giving it a three-dimensional effect. In some cases playfulness cannot be denied (at the Rouffignac cave a horse's head is painted on a flint nodule and when turned ninety degrees the nodule itself looks like a horse), demonstrating humor, artistry and creativity all in one.

In reading this book, you will no doubt want to know the meaning of cave art? The goal here is not to present a "meaning," of cave art but rather to show that this art was motivated by climate change, hence the title of the book. Climate change is in the news today, but Paleolithic people (*homo sapiens*) survived through the most severe swings in climate this planet has experienced in the past two million years; our cousins the Neanderthals, on the other hand, did not make it through this period, possibly because of the extreme climate changes. The idea that Franco-Cantabrian cave imagery accurately reflects the climate at that time based on the animals depicted does not agree with most of the cave art literature today going back to the nineteenth century. In this regard, this book is outside the mainstream. A central tenet of cave art research is that the painters did not represent what was in their immediate environment. In other words, the animals depicted do not represent visual reality. I call this the Cave Art Fallacy, and it is discussed in chapter 7. The Cave Art Fallacy allows the climate hypothesis at first to be entertained, and then to be tested (see chapter 8).

The spark of the idea for this book came during a family reunion in western Ireland, several years ago, but the seeds of the idea grew the following year, during a trip to see some well-known Paleolithic cave paintings in the Dordogne region of southwest France. The previous year in the west of Ireland, during a day trip to the Burren, a region in County Clare, we visited a megalithic tomb called Poulnabrone ("hole of sorrows," *brón* being Gaelic for sorrow), which is a simple rock structure, about six feet in height, composed of a large horizontal flag held aloft by several similarly shaped vertical limestone flags. Archaeological excavations revealed that the burial chamber contained the disarticulated bones of about sixteen adults (men and women) and six children. It was estimated that most of the adults died before the age of thirty, with one individual living to forty. Wear of the vertebrate bones indicated a life of hard physical labor, and dental data suggested a diet of ground cereals. Artifacts found in the burial chamber included stone axes, scrapers, arrowheads, a stone pendant, and two quartz crystals.

In the same field, not far away from the tomb, I noticed a conspicuous large pink granite boulder, sitting on the gray bedrock limestone that is typical of the Burren region of County Clare—and indeed of all of the midlands of Ireland. These conspicuous boulders are scattered throughout much of northern County Clare. I recognized the source of these boulders as Galway granite. Galway is situated only 60 km (37 mi) to the north. Some as big as an automobile, these boulders were left behind after the ice sheets retreated north, about 11,500 years ago, at the end of the last Ice Age.

The glacial erratic and the stone tomb together in the same field led me to ponder what later turned out to be a rather naive question: how did the Stone Age people who built the tomb respond to or cope with the rigors of ice-age conditions? On returning to Dublin, at the National Museum, I asked one of the curators about the age of the stone tomb. Similar Neolithic tombs are scattered throughout Ireland, she replied, and Poulnabrone was between 5,000 and 6,000 years old, according to carbon-14 dating. This was much younger than I had guessed. So, these Neolithic farmers knew nothing about harsh ice-age conditions because they lived over 5,000 years after the last glaciers had disappeared—and the disappearance of the glaciers ushered in a period of warm and stable climate that continues up until today. This stable warm period is so distinctive it is called the Holocene, meaning entirely recent, epoch. The Neolithic people who erected the stone tombs in the Burren region and elsewhere in Ireland lived and farmed in a relatively mild, hospitable climate.

The following summer, I visited a friend in the Languedoc region of southern France and decided to take a side trip to visit some Paleolithic cave paintings for which southwest France is famous. I took a train to Brive-la-Gaillarde, a beautiful midsized town in the Dordogne region in southwest France. My plan was to visit two closely located caves the first day—Lascaux and Rouffignac—which are only 30 km apart. The cave at Lascaux was a facsimile of the original cave. It faithfully reproduces the original cave (called Lascaux II). The nearby original cave was sealed in 1963 to protect the paintings from fungi brought in by tourists. The walls of Lascaux are dominated by painted horses, red deer, mountain goats, and large aurochs (an extinct type of wild cattle) painted in black and various shades of red and yellow. The impressive, very large aurochs, or bulls, dominate the entrance to the cave, and they are the largest images in all of Franco-Cantabrian cave imagery.

After completing my visit at Lascaux, I made a short drive across the valley and arrived at the enormous Rouffignac cave, which is an original cave. Here, the visitor travels inside the cave in relative comfort on a small rickety train. As one train departs from the cave entrance for the interior, a second train arrives on a return journey out of the darkness, filled with beaming tourists. The contrast between Rouffignac and Lascaux caves could not be greater. The animals depicted at Rouffignac are dominated by woolly mammoth, with an occasional woolly rhinoceros and mountain goat (or ibex), and some bison along the walls—there are 158 woolly mammoths, all told. The question that immediately arises is: why do horses and aurochs dominate at Lascaux, whereas the woolly mammoth dominates the cave at Rouffignac, only 30 km (20 mi) away? These paintings are thought to have been painted *during* the late Pleistocene Ice Age.

This rekindled my original question from Ireland the previous year and added some new ones: How did the Upper Paleolithic painters and hunters respond to climate change during the Ice Age? Why are Lascaux and Rouffignac so different? How many millions of tourists have asked themselves the same question? Curiously, these questions have not been answered by cave art specialists, and indeed, they have rarely been asked. This book provides answers to these and other questions. A map of the locations of caves referred to in the text is provided at the end of the book. I thank Jean Clottes for comments on an earlier version of the manuscript.

Kieran O'Hara,
Lexington,
Kentucky,
USA.
August 2014

Introduction

During the 1850s, geologists such as Louis Agassiz established that there was an Ice Age in the Earth's recent past in which ice sheets had covered northern Europe and the Alps. Agassiz's ideas soon took hold in North America, where similar evidence for Ice Age glaciers existed. The study of this period of Earth history opened up a new science—paleoclimatology. At about the same time, archaeologists were discovering that Stone Age man in Upper Paleolithic time had produced sophisticated portable artworks, what the French call *art mobilier*. These include small engraved or painted objects made of bone, ivory, antler, or irregular pieces of slate called *plaquettes*, commonly engraved with depictions of animals. Cave art of the same age was discovered shortly after the recognition of portable artworks. The earliest cave art, thought to have been Upper Paleolithic in age, is Chabot, in the department of Gard, southeast France. It was discovered in 1878. The walls of that cave were decorated with the now-extinct woolly mammoth. The two new disciplines—the study of prehistoric art and paleoclimatology—progressed in parallel at about the same time, and the disciplines would inform one another.

For example, excavations at Kesslerloch, Switzerland, in 1873 indicated that the animal bones at the cave site, representing diet, were mainly reindeer and Alpine hare but also included woolly mammoths and woolly rhinoceros, suggesting an arctic climate.[1] This was consistent with what the geologists were learning from the study of glacial deposits left after the glaciers melted. At the Kesslerloch site, portable art included fine depictions of horses and reindeer. A natural question that arises is whether climate played an important role in what the Upper Paleolithic artists painted during the Ice Age.

By the time that Lascaux was discovered in 1940 and Rouffignac in 1956, archaeologists had already concluded that climate did not play an important role in this artwork because the animals depicted on both portable artifacts and in the caves did not correspond to the bones present at those sites; therefore, the archaeologists concluded that the artwork did not reflect what they hunted or ate.[2] In other words, portable art and cave art were not accurate representations of the immediate environment. The chief example of this mismatch was, and still is, Lascaux. At Lascaux, 88 percent of the bones found in the cave were of reindeer, indicating this was an important part of the diet, but the animals depicted in the cave are predominantly horses, aurochs (extinct cattle), and red deer, with only one reindeer depicted.[3] These results and similar results at other sites led to the same conclusion: the artists did not paint what they hunted.[2,4,5] A corollary of this conclusion is that they did not paint what was in their surrounding environment, and therefore the animal depictions did not reflect the climate at the time. From the outset, theories of cave art involving selective depiction of certain animals for cultural, religious, magical, or ritual purposes therefore gained favor over theories of visual realism. The theory of hunting magic, which was popular until the 1950s, did however involve visual realism (see below).

The underlying assumption in all of this is that the kitchen refuse bones are the same age as the cave artwork—but this has rarely, if ever, been demonstrated. The Lascaux cave was poorly excavated, and much of the archaeological record was destroyed by workmen clearing the cave for tourists in the 1940s and 1950s.[6,7] The French prehistorian Jean Clottes noted that "in most of the deep caves, archaeological evidence was destroyed as they were discovered."[8] The cave painters nevertheless often portrayed the animals they saw with great fidelity and realism, including seasonal coat coloring, detailed musculature, and gender differentiation.[9] Reconciling these conflicting views, namely visual reality on the one hand, and animal imagery that is not representative of the environment on the other, has not been achieved and has been largely ignored.

Whatever the motive for portraying these animals, in a rapidly changing climate typical of the last Ice Age (see chapter 3), the composition of herds encountered by the hunters would have changed with the climate. André Leroi-Gourhan, a French archaeologist, in an exhaustive study of cave art for that time,[10] noted that the horse and bison was by far the most common pair of animals in cave art, and he was puzzled when woolly mammoths dominated in caves such as Rouffignac and Chabot. The Rouffignac cave was never inhabited, so there are no mammoth bones present. The issue of the preponderance of mammoths in some caves is taken up again in chapter 9.

The influential book on Paleolithic cave art and portable art entitled *Journey Through the Ice Age*, published in 1988 and 1997, does not discuss climate.[2] The misleading title may be understood in the context that it is only in the past two or three decades that detailed climate records have become available, whether from the study of cores of deep ocean floor sediments or ice cores retrieved from Greenland and Antarctica. It is here that the science of paleoclimatology can inform cave art prehistory.

Before exploring the issues of Ice Age climate and cave art further, it is best to first establish the time framework we will be dealing with in this book: It spans the interval referred to as the Upper Paleolithic, from about 40,000 years ago to about 11,000 years ago (figure 1.1). The best records of climate over this period come from ice cores retrieved from Greenland. In arctic regions, as the ice accumulates, air is trapped as bubbles in the ice, and

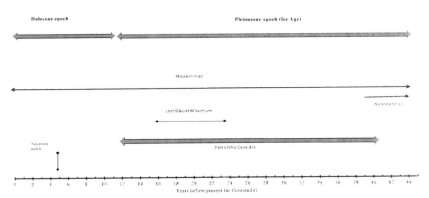

Figure 1.1: A timescale for the last 45,000 years showing two geologic epochs and the overlap of modern man (*Homo sapiens*) with Neanderthals. The Last Glacial Maximum (LGM) and the time span of Upper Paleolithic cave art are indicated. The Neolithic tomb refers to Poulnabrone, Ireland (see Preface). Ages are in calibrated years.

these bubbles represent samples of the atmosphere at the time the ice accumulated. During the last Ice Age, the concentrations of the greenhouse gases carbon dioxide and methane fluctuated in concert, with high concentrations of these gases indicating warm periods, and low concentrations indicating colder periods. Because ice accumulates on an annual basis, ice layers can be individually counted, providing an accurate timescale in years. This ice timescale is more accurate than estimates based on archaeological stratification alone and also sometimes more accurate than the best carbon-14 dates.

Two geological epochs are shown in figure 1.1: The Holocene (a period of warm climate beginning about 11,500 years ago and lasting until today) and the older Pleistocene epoch (corresponding to periods of cold-dry climate alternating with warmer-wetter climate), which began as much as two million years ago. The most recent and intense part of the Ice Age is called the Last Glacial Maximum (LGM); this is when methane (and carbon dioxide) levels in the atmosphere were at their lowest levels and the climate was cold and arid, as much as 20°C (36°F) colder than today's average Mediterranean temperature.[11] This period lasted from about 25,000 years ago to about 15,000 years ago. European cave art shows a major hiatus about 26,000 to 23,000 years ago, when it is thought that the caves may have been too cold to decorate.[12] The human population at that time in Europe was largely confined to refuges in northern Spain and southwest France.[13] Before the LGM (prior to 25,000 years ago), the ice-age climate was very unstable, compared with today's climate, showing several warmer interglacial and colder glacial periods on a millennial scale going back at least 800,000 years, based on ice cores from Antarctica (see chapter 3). The end of the Ice Age marks the beginning of the Holocene epoch, referred to as the Mesolithic and Neolithic periods by archaeologists—the Neolithic stone tomb from County Clare referred to in the preface, which is between 5,000 and 6,000 years old, long after the Upper Paleolithic cave painters had disappeared, is also indicated on figure 1.1 so as to put these Neolithic farmers into context.

Modern man (or *homo sapiens*) arrived in Europe about 40 to 45 thousand years ago from Africa, via the Near East, and overlapped both geographically and for a relatively short period, chronologically, with the Neanderthals, who had been in Europe already for at least 200,000 years, but who retreated to southern Spain before becoming extinct about 40,000 years ago for reasons that are still debated (see chapter 2). The two groups probably overlapped in time for a period of several thousand years.

In this book, I argue that climate was a major factor in what the Upper Paleolithic cave artists depicted and that climate by itself can account for two-thirds of the variation observed in the large animals depicted in French and Spanish cave artwork (see chapter 5). In chapter 7, I show that the apparent mismatch between cave imagery and bone remains is to be expected, and it is an artifact of rapid climate change together with the uncertainty associated with carbon-14 dates. In short, it is the result of comparing archaeological bone remains with cave imagery of a different age. An additional argument for climate playing a major role is that directly dated cave art, when it is classified as dominated by cold-adapted versus temperate animal species, plots on the Greenland ice core climate curve. The chance of this occurring by accident is quite small (chapter 8).

To place this climate-based hypothesis in the context of traditional hypotheses for cave art, a brief overview of the major theories of cart art is outlined below. Upper Paleolithic cave art shows a coherent unity over a period of about 25,000 years (from Chauvet in the Ardèche, one of the oldest decorated caves, to Le Portel, Ariège, France, one of the youngest). Its characteristics are that relatively few animal species are depicted, and they are usually drawn in profile, with indifference to scale and without an environmental context (e.g. ground level or horizon); there are few plant or human depictions apart from hand stencils; the images are usually engraved or painted in either black or red, the primary colors of the painters. Broadly speaking, theories of this cave art fall into the following categories[2,14]:

1. Art for art's sake[15]
2. Sympathetic magic rituals to ensure success in hunting[16]
3. Gender-based French structuralism interpretation[10,17]
4. Magico-religious rituals (e.g., shamanism, totemism)[8]
5. Celebration of hunting itself by young males[9]
6. Information transfer between groups[18]
7. Nihilism

The first theory above dates back to the nineteenth century and was based on portable art, before Paleolithic cave art was discovered. The idea was that there was no intrinsic reason for the artwork other than aesthetic reasons in a time of plenty—in other words, creating art was essentially a leisure activity. As more and more cave art was discovered and documented, it became clear that much cave art was located in isolated, dark regions in the interior of caves, and so it was unlikely to be a leisure activity and could not have been undertaken for ostentation or aesthetic reasons. The second theory, first put forward in 1903 by S. Reinach, relates to success of the hunt and suggests that by representing the hunted animals on the cave walls, the hunter gained power over the animal. About 15 percent of bison are shown speared by javelins[10], a large percentage, considering that a hunter would only rarely take a shot unless success was nearly guaranteed. This theory is consistent with the fact that most cave art deals with large animals that were potential food sources. The hunting magic theory is not inconsistent with the observation that the paintings were commonly confined to the less-accessible parts of the caves where rituals may have taken place. That Australian aboriginals and Native American Indians also believed in hunting magic has been used to support this idea.

The fourth group of theories, partly based on ethnographic analogy to modern cultures in the arctic regions and to South African tribes, relates to totemism and shamanism.[8] Shamans entered a trance-like state in some societies and were consulted about the future, the weather, and other matters. The fifth theory, that cave art was a celebration of hunting, especially by young male hunters showing off their hunting prowess, has been documented in the book entitled *The Nature of Paleolithic Art*.[9] The visual realism in many cases of the animals portrayed, including details of the seasonal variation in the animal's coat, gender differentiation, and detailed musculature of the animals is consistent with this idea. The sixth theory above, that cave art was a form of information transfer across generations and between widely separated hunting groups during periods of rapid climate change is plausible.[18] This theory is also consistent with the work of R. D. Guthrie[9], insofar as he has documented the visual realism with which the large animals were portrayed by the cave painters. Guthrie also emphasizes the astute observational acumen of the hunters based on the details in the drawings. Theories of cave art are taken up again in the final chapter.

The last theory, nihilism, of course, is not a theory at all. It posits that there is no point in trying to understand the meaning of cave art because it is unknowable.[19] This is an unscientific (if not anti-scientific) viewpoint. Normally in science one gathers information and evidence and then follows up with a hypothesis to explain the facts. One vocal pundit tends to deride other theories but has failed to produce his own.[20]

The main goal of this book is not to present a "meaning" for cave imagery, but rather to show that climate change during the late Pleistocene epoch was the main motivation behind it. In the end, the cave painters did not run out of paint or caves; rather, the climate stabilized at the end of the Ice Age, thereby removing its raison d'être. How and why cave art lasted so long is addressed in the final chapter. Before examining the Ice Age climate in more detail (in chapter 3), we first address the question, in the next chapter, of who the cave painters were.

References cited

1. Merck, K., 1876. *Excavations at the Kesslerloch site near Thayngen, Switzerland.* Longmans, London.
2. Bahn, P. G. and Vertut, J. 1997. *Journey Through the Ice Age.* University of California Press.
3. Bouchud, J., 1979. *La faune de la grotte de Lascaux in Lascaux Inconnu*, Leroi-Gourhan, Arl. & Allain, J (eds.), 147–152, CNRS, Paris.
4. Delluc, B. and Delluc, G. 1984. "Le tableau de chasse." *Histoire et Archeologie*, 87, 28–29.
5. Altuna, J. 1983. *On the relationship between archaeo-faunas and parietal art in the caves of the Cantabrian region in Animals and Archaeology: Hunters and their Prey.* Clutton-Brock, J. and Grigson, C. eds., 1: 227–238.
6. Leroi-Gourhan, Arl., 1979. *La stratigraphie et les fouilles de la grotte de Lascaux in Lascaux Inconnu.* Leroi-Gourhan, Arl and Allain, J. (eds.) 45-74, CNRS, Paris.
7. Delluc, B. and Delluc, G., 2006. *Discovering Lascaux.* (English edition), Éditions SudOuest.
8. Clottes, J. and Lewis-Williams, D., 1998. *The shamans of prehistory: Trance and magic in the painted caves.* Abrams, New York.
9. Guthrie, R. D., 2005. *The nature of Paleolithic art.* University of Chicago Press.
10. Leroi-Gourhan, A., 1967. *Treasures of Prehistoric art.* Abrams, New York.
11. Renssen, H. and Isarin, R.F. B., 2001. "The two major warming phases of the last deglaciation at 14.7 and ~11.5 ka cal BP in Europe: climate reconstructions and AGCM experiments." *Global and Planetary Change*, 30, 117–153.
12. D'Errico, F. et al., 2001. "Les possible relations entre l'art des caverns et la variabilité climatique rapide de la dernière période glaciaire" in Barrandon, J-N. et al., eds., *21st Rencontres Internationales d'archéologie et d'histoire d'Antibes*, Éditions APDCA, Antibes, 333–347.

13. Bocquet-Appel, J-P. and Demars, P-Y., 2000. "Population kinetics in the Upper Paleolithic in Western Europe." *J. Archaeological Science*, 27, 551–570.
14. Lawson, A. J., 2012. *Painted caves. Paleolithic rock art in Western Europe*, Oxford University Press.
15. Halverston, J. 1987. "Art for art's sake in the Paleolithic." *Current Anthropology*, 28, 63–89.
16. Reinach, S. 1903. "L'art et la Magie. A propos des peintures et des gravures de l'age du Renne." *L'Anthropologie*, 14.
17. Ucko, P. J. and Rosenfeld, A., 1967. *Paleolithic cave art*. World University Library.
18. Mithen, S. J., 1988. "Looking and learning: Upper Paleolithic art and information Gathering." *World Archaeology*, 19, 297–327
19. Thurman, J., 2008. "First Impressions. What does the world's oldest cave art say about us?" *The New Yorker*, May 2008.
20. Bahn, P. G., 1997. "Membrane and numb brain: a close look at a recent claim for shamanism in Palaeolithic art". Rock Art Research, 14, 62-68.

2 Who Were the Cave Painters?

Archaeologists divide the latter part of the Pleistocene epoch, a period referred to as the Upper Paleolithic, into several distinct cultures (also called traditions, technologies, or industries) based on the archaeological record of stone tools and other artifacts made of bone, antler, or ivory.[1] From youngest to oldest, these technologies are named after their type localities: Aurignac, La Gravette, Solutré, and La Madeleine in France. They are Aurignacian, Gravettian, Solutrean, and Magdalenian traditions. For the interested reader, *Cave Art* by Jean Clottes illustrates cave art and portable art from each of these different cultural periods.[2]

Table 2.1 below presents the time frame for the different cultures associated with cave art. The first column of dates shows raw radiocarbon (carbon-14) ages in thousands of years (ka) and is only approximate. The radiocarbon timescale is non-linear and needs to be stretched in places (calibrated is the correct word), especially in the interval 10,000 to 50,000 years before the present, because of variations in the production of carbon-14 in the atmosphere over time (see chapter 6). This natural variation in carbon-14 production was not recognized early on in the development of this dating technique (1950s), so that earlier dates were inaccurate to some degree, leading to difficulty with correlating climate events with archaeological events. For example, the Chauvet cave art (in Ardèche, southeast France), yields raw radiocarbon dates of about 32,000 in carbon-14 years before present, but when calibrated, the corrected age is 36,000 calendar years before present, which is a large correction.[3] The correction for older ages is more uncertain. The second column of figures in table 2.1 represents calibrated years using the most recent calibration curve.[4] In this book, calibrated dates are expressed as ka cal BP, indicating calibrated thousands of years (ka) before the present (BP). Uncalibrated dates are indicated as ka ^{14}C BP. Dating methods and calibrations are discussed in chapter 6.

Table 2.1 Upper Paleolithic Cultures

Technology	^{14}C years	Calibrated years[4]
Magdalenian	18-11 ka*	21 - 13 ka
Solutrean	21-18 ka	26 - 21 ka
Gravettian	29-21 ka	35 - 26 ka
Aurignacian	35-29 ka	42 – 35 ka

* Thousands of years before present

The four cultural periods spanning the period of cave art using calibrated years is shown in figure 2.1, together with other characteristics of those cultures. This figure will put some of the events discussed below in context and help explain who the cave painters were, bearing in mind that these cultures were likely time-transgressive, having different time spans

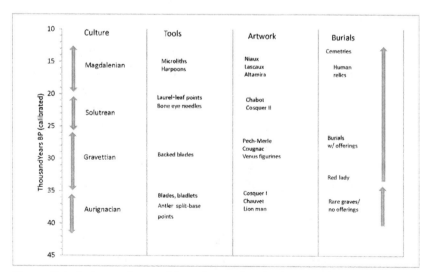

Figure 2.1: A time scale in calibrated years for the four cultural periods of the Upper Paleolithic. Some of the characteristics of these cultures are indicated.

in different locations. It is also worth noting that the Magdalenian time period, for example, is separated in time from the Aurignacian by the same amount of time as we are separated from the Magdalenians (about 15,000 years).

Who painted the caves? Since the Neanderthals became extinct about 40,000 years ago but cave painting has a certain unity throughout (restricted number of species, animals in profile, without environmental context or scale) which continued up until the end of the Magdalenian period, cave paintings have been almost exclusively associated with modern man (*Homo sapiens*), people who were cognitively similar to ourselves. Genetic studies of modern Europeans compared with the DNA in Neanderthal bones indicate we have little (a few percent perhaps) of their genetics, and little if any interbreeding appears to have occurred between our ancestors (*Homo sapiens*) and the Neanderthals.[5] Some scholars, however, argue that Neanderthals developed their own advanced culture (recorded in Châtelperronian deposits, a cultural period that predates the Aurignacian), including evidence for bodily decoration, jewelry, and advanced stone technologies, suggesting they were cognitively as advanced as modern man (*Homo sapiens*).[6,7]

Other scholars counter that the Neanderthals only copied from the newly arrived immigrants (modern humans) by exchange and acculturation and did not have the capacity themselves to develop their own symbolic culture. Symbolic thought is expressed in such things as jewelry (e.g., perforated animal teeth used as pendants), decorated artifacts of bone and antler, burial graves associated with valuable personal objects, or bodies decorated with pigment.

More recent studies, however, suggest that the stratigraphy of several Neanderthal sites has been disturbed so that fossil remains and artifacts have been mixed, but that they originally came from different stratigraphic levels so that their coexistence in time cannot be demonstrated.[8] This means that the Neanderthal fossil remains and the artifacts cannot be reliably associated with each other. This is a common feature of archaeological sites in Europe that were either improperly excavated in the nineteenth century or were disturbed by the original inhabitants themselves or by animals, such as bears or hyenas, that entered the caves and disturbed bones or artifacts. Moreover, recent radiometric determinations using the latest chemical and statistical techniques at the type Châtelperronian locality (in Grotte du Renne, France)[9] produces reversals, whereby the relative stratigraphic age and the radiometric ages are inconsistent—a feature consistent with stratigraphic levels that have been physically mixed by any of the processes just mentioned. This makes claims of Neanderthal intellectual equality difficult to sustain, but the issue, like many in archaeology and anthropology, is hotly debated.[10]

Recently, a red hand stencil from a cave in northern Spain, El Castillo, was dated to about 41 cal ka BP using the uranium-thorium dating technique; thus it represents some of the oldest cave art in Europe.[11] Negative hand stencils are thought to have been produced by blowing paint (red ocher mixed with water) from the mouth or through a straw onto the cave wall, and clearly it is not a particularly sophisticated form of cave art. Because Neanderthals were extant in northern Spain at that time, the authors of the study stated that it cannot be ruled out that the stencils were painted by the Neanderthals. This statement drew a sharp response from French archaeologist and cave art scholar Jean Clottes, who termed the reference to Neanderthals as "gratuitous" and said that it was clearly only designed to get popular media attention, which it did.[12] The El Castillo paper was also criticized on the grounds that, in limestone caves, uranium can be mobile, which would increase the thorium/uranium ratio, making the sample appear older than it really is. There is no archaeological evidence that Neanderthals ever engaged in cave art, even if it was only hand stenciling, although as mentioned already, several workers believe they were capable of it.[6,7]

Modern humans first appeared in Europe, from sub-Saharan Africa, via the Near East, about 40,000 to 45,000 years ago and gave rise to the Aurignacian cultural tradition[13,14]; this means that the first cave painters were probably Aurignacians. That some of the oldest cave art is also the most sophisticated (e.g., the Chauvet cave) undermines the traditional view that the cave artists increased in skill and sophistication over a long period of time, up through the Magdalenian cultural period. Chauvet and Lascaux are shown in figure 2.1, underscoring the large time interval separating the two, although in terms of sophistication of the artwork they are comparable.

There are two methods of dating cave art: One is based on stylistic considerations where certain characteristics of the drawing, depiction of horns in frontal view or profile view or the outline of the mammoth's body for example, are thought to be characteristic of that time period, in addition to such techniques as perspective or the dynamic postures of the animals. The French archaeologist and cave art scholar Leroi-Gourhan laid the foundations of this method of stylistic dating, although Abbé Breuil also made stylistic determinations much earlier in his book *Four Hundred Centuries of Cave Art*. These determinations are relative in that one could say, for example, that this cave art was lower Magdalenian in age, but some other cave art was middle Magdalenian, so that the latter would be younger. The second method of dating is based on radiometric dating techniques, such as the carbon-14 method, where specific ages in years (with associated statistical error) can be assigned to the paintings (see chapter 6).

The Aurignacian age of Chauvet (based on radiometric ages), together with the sophistication of the paintings, which, based on stylistic considerations, is considered to be Solutrean or Magdalenian in age, left some scholars perplexed.[15,16] They refused to believe the chronometric evidence, even as the number of radiometric ages approached 100, from ten different laboratories, yielding similar results indicating an Aurignacian age.[17,18] Although several wall paintings were dated directly at Chauvet by sampling carbon from the paintings themselves, most of the 100 dates came from charcoal in cave hearths and yield the same age as the nearby paintings. The only possible way to explain why these charcoal dates would not reflect the age of the paintings would be to hypothesize that a group of cave painters entered the cave at a much later date and used old charcoal that was already in the cave. But in that case, why the pieces of charcoal were produced in the first place must be explained because the charcoal was not simply for heat or lighting but appears to have been produced for the purpose of painting, on account of the large size of the charcoal pieces.[17] The four dates from the paintings themselves must then also be explained away as being erroneous for technical reasons.[19] Disagreements between radiometric ages and stylistic age determinations on cave art are quite common, but sometimes they do agree, which makes everybody happy.

There is much at stake here, however, in the case of Chauvet—not just the age of a single cave's artwork. When did hominins, for example, develop superior cognitive abilities, including language and abstract symbolic thought—what is referred to as modern behavior. Could it predate modern man's arrival in Europe, and could the Neanderthals also have been affected by these crucial advances?[20] Chauvet would seem to demand that the Aurignacians either were already skilled cave painters when they arrived in Europe or that they learned their new craft very rapidly, within a few thousand years of arrival. The puzzle of Chauvet is that it appears largely out of context, although other Aurignacian cave art is now recognized in northern Spain[21] and sophisticated portable art of Aurignacian age from southern Germany (for example, the man with a lion's head) has been known for some time.[2] A cave close to Chauvet, La Baume Latrone, is now also thought to be Aurignacian in age although not as accomplished as the artwork at Chauvet. As one expert noted, finding Chauvet was similar to finding a Renaissance masterpiece in a Roman villa.

Why the Neanderthals became extinct about 40,000 years ago to be replaced by modern humans is debated, but the two most favored theories are deteriorating climate conditions together with competition from the new arrivals.[5,22] Since the Neanderthals had already become acclimatized to cold conditions for at least 200,000 years in Europe, it may seem counterintuitive that they lost out to the new arrivals, who were not only unaccustomed to cold climate but who came from a subtropical African climate, via the Near East. It appears that the technological superiority of *Homo sapiens* played a role. The population of the new arrivals increased tenfold as the population of the existing Neanderthals decreased.[23] At least one researcher has speculated that the ability of the new arrivals to produce more advanced clothing involving closely fitting skins was significant. This presumably required sewing hides together, possibly in double layers, and fastening them with buttons or pegs, allowing the wearer to hunt in colder conditions.[24] In contrast, the Neanderthals may have had only a single layer or wrap-around clothing, which did not involve sophisticated tailoring or sewing. All evidence of their clothing, however, is lost.

We do know what kind of clothes a later Neolithic hunter-herdsman wore. Ötzi the ice man was found in a glacier in the Italian Alps in 1991 and is dated to about 5,300 years old; some of his clothing was preserved.[25] The man himself was about thirty-five or forty years old. He wore leather shoes, the soles of which were thick and made from bear skin, and the uppers were made from red deer. His legs were covered in individual leggings, which were carefully sewed together from smaller pieces of hide but were not attached to one another as in a pair of trousers. He wore a loincloth kept in place by a belt around his waist. This strap apparently also kept the leggings in place, attached to suspenders. The leggings were tucked into his shoes, and the shoes were insulated with grass or straw. A leather hat was found, also made from bear skin. His upper torso garment was not well-preserved but appeared to be a coat-like fur skin, kept closed by a belt at the waist. Over these clothes, he wore a woven grass cape that came down to his knees to protect him from the rain.

The first appearance of eyed bone needles, suggesting sewing, is not until the Solutrean, well after the Neanderthals' demise, but small, round, perforated bone objects that could serve as buttons are common in the Aurignacian and earlier periods. Also, that the last Neanderthals are found in southern Spain suggests that they took refuge southward during cold periods and were no longer able to compete for resources with their technologically more advanced cousins.

In general, the cave painters did not dwell in or inhabit the caves they painted; they only visited them for the purpose of painting and possibly ritual activities. Caves are dangerous places—they are prone to flooding and wild bears hibernate there (as indicated by bear scratch marks on many cave paintings and hollows on the floor in which they slept). Also, one wrong footstep in a dark limestone cave, which commonly has several levels connected by vertical shafts, could lead to serious injury or death. The cave painters were hunter-gatherers, and they were mobile, following herd migrations on a seasonal basis, and they lived either in rock shelters near the front of caves or in mobile encampments. Some of the more salient features of the four main cultural periods of the Upper Paleolithic are briefly described below.

Aurignacian (42–35 cal ka BP)—The Aurignacian lithic technology is characterized by parallel-sided blades (two to three cm in width), and smaller blades or "bladelets" (less than one cm wide), fashioned from larger, pyramid-shaped cores of rock, such as flint, by directed blows to the core. These tools were commonly maintained and sharpened by retouching. This technology is also associated with split-base points made from reindeer antler—the split-base indicating that the points were probably mounted onto wooden shafts and possibly used as javelins in big-game hunting. A study of the provenance of Aurignacian stone tools in the Dordogne region of southwest France gives some idea of the mobility of these early hunter–gatherers.[26] They commonly set up temporary hunting camps for the fall migration of animal herds. They used a variety of silica-rich rocks as source materials for their tools, such as chert, flint or obsidian (a volcanic glass). These materials' distinctive colors and natural markings allow archeologists to locate their source, whether in local river beds or rock outcrops in the region. These studies show that most of the material for stone tools, about 60 percent, was sourced less than 5 km (3 mi) away. The relatively lower percentages of materials from distances of at 35 km and 40 km, on the other hand, correspond to localities where a type of Cretaceous chert that was highly prized and suitable for certain types of tools such as spear blades and smaller flat bladelets. These materials showed almost 100 percent usage at the campsites, with little waste, whereas large amounts of local materials often went unused.[26] This would suggest that traveling long distances was not taken lightly, and when they needed such materials, they were going to use them fully. Five kilometers is thought to have been a typical foraging distance for these hunters. More recent studies of Magdalenian encampments indicate these hunter-gatherers traveled significantly greater distances for their resources, 50 km to 100 km (see below).

Gravettian (35–26 ka cal BP)—The subsequent Gravettian cultural period shows evidence for the burial practices of Upper Paleolithic man.[27] A geologist, William Buckland, in 1823 described a human burial in a cave in Wales overlooking the Severn estuary. The burial turned out to be the first known Upper Paleolithic burial. Because the skeleton was stained in red ocher, the burial was dubbed the "Red Lady," although it was subsequently identified as an adult male. It seems likely that the body was clothed at the time of burial, the clothing itself being stained with ocher, although nothing remains of the clothing. Grave offerings included perforated periwinkle shells near the thigh bone (possibly originally in a pocket), and several short ivory rods and rings, which were also stained red and placed on and around the body. A complete mammoth skull was found close to the skeleton, and it has been suggested that this was also part of the grave offerings. Radiocarbon dating of the skeleton yielded an age of 28 to 29 ka ^{14}C BP (33–34 ka cal BP), placing the burial in the early Gravettian cultural period (the Red Lady is indicated on figure 2.1).

Burials of this period in Europe are common and show very similar patterns of grave offerings involving perforated teeth, shells, ivory objects, and pigment staining. The earlier Aurignacian period does not preserve evidence of this type of symbolism associated with burial, and neither do Neanderthal sites show grave offerings.[27] It appears that this type of burial only became common after about 35,000 years ago. The grave offerings imply that the Gravettians had some ideas about an afterlife, suggesting the ability for abstract thought. Decorated objects and personal jewelry and grave offerings all suggest that these people were capable of symbolic or metaphoric thought. Many scholars believe that this type of mental activity also indicates capability for complex spoken language.[28] In light of these findings, it should not be too surprising that these people were also capable of elaborate cave paintings, such as the famous polka-dotted horses at Pech Merle and the mammoths at Cougnac in southwest France, or the ivory sculpture called the lady with a hood from Grotte du Pape, France.[2] Portable art of the period is also characterized by the presence of Venus (female) figurines.[2] The lithic technology of the Gravettian period is characterized by "backed," or one-sided, blades used as knives.

Solutrean (26–21 ka cal BP)—The following Solutrean industry displays a distinctive lithic assemblage. These tools, called laurel-leaf points (*feuilles de laurier*), are very finely knapped on both sides and were probably used as weapon tips on spears used in close-range hunting.[1] These stone tools are so similar to the later Clovis points found in North America (circa 13.5 ka years BP) that their discovery led to the provocative "Solutrean hypothesis," which posits that Solutreans from Europe colonized North America by traveling along the southern margin of the ice sheet in boats[29]. But there is no evidence the Solutreans had the technology for such long-distance seafaring journeys; in addition, the timing appears to be off by a large amount.

As mentioned already, one of the first depictions of cave art thought to be Upper Paleolithic in age is the Chabot cave (or rock shelter) discovered in 1878, and it is of Solutrean age. Its age is indicated by excavated deposits and radiometric dates yield ages of 18. 2 ka ± 400 ^{14}C BP and 17.7 ka ± 400 ^{14}C BP[2] (approximately 22.0 ka cal BP). Of the nineteen animals depicted near the entrance to the cave, ten are mammoths, not including several outlines of unfinished mammoths. The Chabot cave dates to the Last Glacial Maximum (LGM), the coldest part the Ice Age.

Magdalenian (21–13 ka cal BP)—The vast majority, about 80 percent, of Upper Paleolithic cave art sites are assigned to the middle and upper Magdalenian, although it should be noted that most of these age assignments are based on stylistic considerations rather than radiometric dates. The reasons for this may be that younger works have a better chance of preservation; in addition Magdalenian cave imagery may have increased due to a large population increase—or the assigned ages maybe incorrect.

The excavation of some well-preserved hunting camps and larger base camps shows that the camps were efficiently organized to perform specific tasks.[30] The site of Pincevent is located 80 km southeast of Paris on the shore of the Seine and was used to hunt reindeer herds during their fall migration. It was repeatedly occupied from about 13.4 to 15.5 ka cal BP (Upper Magdalenian), as indicated by radiocarbon dates on charcoal and bone.[31] The site has been excavated over an area of 4,500 square meters (15,000 square feet), and the evidence indicates that, at any one time, the entire camp was used. Four domestic areas with eighty hearths were identified, and several surrounding open-air areas were used for butchering and stone tool knapping and as waste dumps. A total of seventy-six reindeer were processed at the site. These features indicate that Pincevent was a large base camp. The reindeer at Pincevent were forced to cross the Seine River during their fall (autumn) migration south, and it is here that the hunters could most easily intercept their prey. More recent excavations at Pincevent have revealed an occupation level that indicates that horse were also hunted year round.[31] Hunting of horse, in addition to reindeer, is now recognized as having been important in the Paris basin at several sites.

Verberie is a second Magdalenian open air site, located 60 km north of Paris that was also occupied about 13.9 ka to 15.5 ka cal BP, about the same time as Pincevent. Eight levels indicate repeated occupation of the site, most likely during the fall (autumn) migration season.[31] Reindeer were again the main prey. In one level alone, forty reindeer were killed and butchered. Other species present were arctic squirrel, arctic fox, horse, and two mammoths, suggesting cold conditions. This site was mainly a temporary butchering camp from which meat was transported, as indicated by intact reindeer skeletal remains, a feature not found at Pincevent. In other words, the meat was removed from the skeleton without disarticulating the bones. The spatial arrangement of stone tools and debris indicate five distinct areas: 1) human resting areas, clear of debris; 2) hearths defined by larger stones; 3) stone tool knapping workshops; 4) butchering areas; and 5) hide preparation areas, indicated by abundant stone tool scrapers and blades for cutting.[30] A third Upper Magdalenian site in the Paris basin, Etiolles, indicates both horse and reindeer were hunted.[31]

Procurement distances for raw materials appear to have been greater compared with Aurignacian times, where outcrops of Cretaceous flint were commonly about 50 km distant—and sometimes considerably more—from encampments in the Paris basin. A characteristic tool of the Magdalenian period is finely carved bone harpoons made of reindeer antler indicating that fish or seal was part of the diet. The lithic industry is characterized by microliths of various shapes. The sites described above give substantial insight into the Magdalenian hunter's lifestyle, both in temporary hunting camps (Verberie) and in larger base camps (Pincevent).

The end of the Magdalenian period is marked by very large and rapid climate fluctuations, including the Bølling-Allerød warm periods that lasted about 1000 years, with a cold snap in between (Dryas II), followed by a severe chill, the Younger Dryas, lasting about 1200 years. The Younger Dryas would have pushed both animal and human populations south to refuges in northern Spain and southwestern France. After this time, cave art of the Upper Paleolithic style is not encountered again. The ice retreated for good at the end of the Younger Dryas, about 11,500 years ago, and warm and stable climate marks the beginning of the Holocene geologic epoch. In the next chapter, the Ice Age climate is examined in more detail.

References cited

1. Richet, C. (ed.), 1984. "Les premiers artistes: derniers chasseurs de la Préhistoire." *Dossiers de Histoire et Archeologie*, October 1987, 1–94.
2. Clottes, J., 2008. *Cave Art*. Phaidon Press, London.
3. Bard, E., 2001. "Extending the calibrated radiocarbon record." *Nature*, 292, 2443–2444.
4. Reimer, P. J. et al., 2013. "INTCAL13 and marine 13 radiocarbon age calibration curves 0–50,000 years cal BP." *Radiocarbon*, 55, 1869–1887.
5. Mellars, P., 2004. "Neanderthals and the modern human colonization of Europe." *Nature*, 32, 461–465.
6. D'Errico, F. et al., 1998. "Neanderthal acculturation in Western Europe? A critical review of the evidence and its interpretation." *Current Anthropology*, 39, S1–S44.
7. Zilhão, J., 2006. "Analysis of Aurignacian interstratification at the Châtelperronian type-site and implications for the behavioral modernity of Neanderthals." *Proceedings of the National Academy of Sciences*, 103, 12643–12648.

8. Bar-Yosef, O., 2010. "Who were the makers of the Châtelperronian culture?" *J. Human Evolution*, 59, 586–593.
9. Higham, T. et al., 2010. "Chronology of the Grotte du Renne (France) and implications for the context of ornaments and human remains within the Châtelperronian." *Proceedings of the National Academy of Sciences* 107, 20234–20239.
10. Banks, W. E. et al., 2013. "Revisiting the chronology of the proto-Aurignacian and the early Aurignacian in Europe: A reply to Higham's comments on Banks et al. (2013)." *J. Human Evolution*, 65, 810–817.
11. Pike, A. W. G. et al., 2012. "U-series dating of Paleolithic art in 11 caves in Spain." *Science*, 336, 1409–1413.
12. Clottes, J., 2012. "U-series dating, evolution of art and Neanderthals." *International Newsletter on Rock Art*, No. 64, 1–6.
13. Mellars, P., 2011. "The Earliest modern humans in Europe." *Nature* 479, 483–484.
14. Benazzi, S. et al., 2011. "Early dispersal of modern humans in Europe and the implications for Neanderthal behavior." *Nature*, 479, 525–529.
15. Pettitt, P., 2008. "Art and the Middle-to-Upper Paleolithic transition in Europe: Comments on the archaeological arguments for an early Upper Paleolithic antiquity of the Grotte Chauvet art." *J. Human Evolution*, 55, 908–917.
16. Zuchner, C., 1996. "The Chauvet cave: radiocarbon versus archaeology." *International Newsletter on Rock Art*, No. 13, 25–27.
17. Cuzange, M-T. et al., 2007. "Radiocarbon inter-comparison program for Chauvet cave." *Radiocarbon*, 49, 339–347.
18. Quiles, A. et al., 2014. "Second Radiocarbon inter-comparison program for the Chauvet-Pont d'arc cave, Ardèche, France." *Radiocarbon*, 56, 13–21.
19. Pettitt, P. and Pike, A., 2007. "Dating European Paleolithic cave art: Progress, Prospects, Problems." *J. Archaeological Method and Theory*, 14, 27–45.
20. Milford, H. et al., 2004. "Why not the Neanderthals?" *World Archaeology*, 36, No. 4, 527–546.
21. González-Sainz, C. et al., 2013. "Not only Chauvet: Dating Aurignacian rock art in Altxerri B cave (northern Spain)." *J. Human Evolution*, 65, 457–464.
22. Banks, W. E. et al., 2008. "Neanderthal extinction by competitive exclusion." *Plos One*, 3, 1–8.
23. Mellars, P. and French, J.C., 2011. "Tenfold population increase in Western Europe at the Neanderthal-to-modern human transition." *Science*, 333, 623–627.
24. Gilligan, I, 2007. "Neanderthal extinction and modern human behavior: The role of climate change and clothing." *World Archaeology*, 39, 499–514.

25. Fowler, B., 2000. *Ice Man: Uncovering the life and times of a prehistoric man found in an alpine glacier.* University of Chicago Press.
26. Blades, B., S., 2001. *Aurignacian Lithic Economy. Ecological perspectives from southwestern France.* Kluwer/Plenum, New York.
27. Pettitt, P., 2011. *The Paleolithic origins of human burial.* Routledge Press, New York.
28. Mellars, P., 2005. "The impossible coincidence. A single-species model for the origin of modern human behavior in Europe." *Evolutionary Anthropology,* 14, 12–27.
29. Bradley, B. and Stanford, D., 2004. "The North Atlantic ice-edge corridor: a possible Paleolithic route to the New world." *World Archaeology,* 36, 459–478.
30. Zubrow, E, Audouze, F. and Enloe, J. G. (eds.), 2010. *The Magdalenian Household: Unraveling Domesticity.* State University of New York Press. New York.
31. Debout, G. et al., 2012. "The Magdalenian in the Paris basin: New results." *Quaternary International,* 272–273, 176–190.

3 A Journey Through the Ice Age

The National Ice Core Laboratory in Denver, Colorado, is operated by the United States Geological Survey, and its purpose is to store ice cores, drilled and extracted mainly from the polar regions and transported to Denver in refrigerated cargo planes and trucks. The facility is held at -36°C (-33°F), which is the optimum temperature to preserve the ice and the gases trapped in the ice. Visitors are given heavy parkas and thick mittens for a tour of the facility. The laboratory houses 17 km (10.5 mi) of ice core from thirty-four drilling sites, largely from Greenland and Antarctica and other glacial regions. The ice cores themselves are about the diameter of a liter-and-a-half wine bottle. They are sliced in half lengthwise, and one half is used for analyses and the other half is stored for future study. Since the ice collections represent a great deal of effort and expense on the part of hundreds of technicians and scientists, there are three backup refrigeration systems at the Denver facility in case of a power failure—that's more backup systems than the Fukushima nuclear power plant had at the time of its accident in 2011.

In Polar regions, each year the snow gets buried under more snow, and eventually it turns into ice due to the overlying weight. Some of the air in the snow gets trapped in the ice as small bubbles. The pockets of trapped air represent small samples of the atmosphere, and they record the chemical composition of the atmosphere at that time. The ice also records layers of ash from far-away volcanic eruptions and also dust blown in from arid regions. The annual layers in the ice can be individually counted, providing an accurate timescale. Some ice cores go back hundreds of thousands of years and provide a very detailed record of past climate during the last Ice Age. A complication to the simple counting of annual layers in the ice to provide a timescale is that there is some difference between the age of the ice and the age of the trapped gases. The gases tend to percolate to deeper levels before pore spaces in the ice are sealed off. This means that the gases are younger than the surrounding ice layers, in some cases by as much as several hundred years, depending on the depth in the core and the snow accumulation rate at that location. The age of the gases can be corrected, however, using mathematical models based on how fast the gases diffuse downward. The resulting timescales of ice cores are sometimes more accurate when compared with radiometric ages (carbon-14 ages, for example). There is a tendency, however, for the accuracy to decrease as one goes farther back in time (i.e. deeper in the core).

The idea that Ice Ages were caused by variations in the Earth's orbit around the sun was developed by James Croll, a nineteenth-century self-taught man of science from Scotland. He also correctly inferred that ocean currents in the Atlantic and the trade winds magnified the relatively small effects of the orbital variations. His ideas on the subject were published in 1885, and they greatly influenced subsequent scientists who tackled the Ice Age problem.[1]

A Yugoslavian engineer, Milutin Milankovitch, joined the faculty at the University of Belgrade in 1909 and turned his interests toward the cause of Ice Ages. He set out to calculate the amount of solar radiation that reached the Earth in the past at different latitudes. He inferred that the amount of sunshine on the continents at high latitudes in the summer would control the advance and retreat of ice sheets. He reasoned as follows: If it becomes colder during the winter, things don't change much because it is already cold. However, if it becomes colder during the summer, less ice will melt, and the following winter, the glaciers will grow larger. Milankovitch set about calculating the changes in heat from the sun due to orbital variations of the Earth around the sun at 45°N latitude in the summertime.[1] All his calculations were done with pen and paper, some of which were done while he was a prisoner for a short time in the Balkans during World War I.

It was already known by Milankovitch that three orbital variations affected the amount of solar heat that reached the Earth over time:

- Variations in the shape of the Earth's orbit around the sun, being more or less elliptical over time. This variation has a periodicity of 100,000 years.
- Variations in the tilt of the Earth's spinning axis. This variation (21–24 degrees) has a periodicity of 41,000 years. This tilt is the cause of our seasons and is currently 23.5°.
- The wobble (or precession) of the Earth's axis, which has a period of 23,000 years.

At certain times, these three periodic changes coincide, causing more or less heating of the Earth by the sun. The results of Milankovitch's calculations, which were a series of sine curves for solar heating over the past 600,000 years, were consistent with what geologists had known about past Ice Ages, but there was no way to prove this. The sedimentary record of past Ice Ages on the continents is fragmentary because, as each new ice sheet advanced over northern Europe and North America, it removed sediments previously laid down by earlier glacial advances. This hindered any attempts to evaluate Milankovitch's cycles. Milankovitch died at seventy-nine, but he was recognized for his important work by advancing Croll's earlier theories.

In the 1970s, climate researchers turned their attention to the deep oceans.[2] Here, the sedimentary record was much better than on the continents and went back several hundred thousand years. A large number of sediment cores were retrieved from ocean basins around the world during a research project called CLIMAP, whose goal was to map the surface temperature of the Earth during the last Ice Age. Within the ocean sedimentary layers, there were small fossils, called foraminifera (forams for short), whose shells were made of calcium carbonate ($CaCO_3$). These animals accumulated at the bottom of the ocean floor when they died, but during their lifetime, they built their skeletons from calcium and oxygen from seawater. One of the main tools used to build a climatic record was a record of the ratio of two types of oxygen atoms (oxygen-18 and oxygen-16) in the skeletons of the fossilized forams (O-18 has two more neutrons in its nucleus than O-16, making it heavier). The resulting curves showed a complicated pattern that reflected the amount of ice on the continents during past Ice Ages—the snow and ice preferentially incorporated the lighter oxygen-16 during evaporation from the ocean, thereby changing the O-18/O-16 ratio. An important paper that resulted from this work was entitled "Variations in the Earth's Orbit: Pacemaker of the Ice Ages.[3]" It examined the periodicity of these curves in two deep-sea cores from the Pacific Ocean and showed that they agreed with the periodicity of Milankovitch's cycles. This was the first strong evidence that orbital variations of the Earth controlled Ice Ages. It also showed that the oceans played an important role in climate change, confirming Croll's earlier ideas.

In addition to the three orbital variations mentioned above, it was known that the composition of the Earth's atmosphere affected the Earth's surface temperature. John Tyndall, an Irish scientist, had shown by experiment in the laboratory in 1859 that carbon dioxide and methane gases trapped heat. These are now known as greenhouse gases because they let in visible solar radiation through the atmosphere, but they block the infrared (invisible) radiation from going back out. Greenhouse gases therefore act as a one-way filter in the atmosphere and cause heating of the Earth's ground surface and atmosphere. The greenhouse effect of methane is twenty-five times more powerful than that of carbon dioxide, but because its concentration in the atmosphere is much less than carbon dioxide, it only produces about 30 percent of the warming compared to carbon dioxide.[4] The Swedish scientist Svante Arrhenius later showed in 1908 that doubling of carbon dioxide in the Earth's atmosphere would result in a global temperature increase of about 5°C (9°F), which is similar to the more modern estimates based on recent computer models.[4] It is now thought that, although greenhouse gases themselves do not initiate or end Ice Ages, their role is to amplify the effects produced by the orbital variations through various feedback mechanisms, which are not yet fully understood.

To study the role of greenhouse gases in climate change, researchers turned their attention in the 1990s to ice cores from Polar regions. One of the first deep ice cores retrieved was from Vostock station in Antarctica. It covered four glacial cycles over the past 420,000 years.[5] The variations of greenhouse gases (methane and carbon dioxide) were not only similar to the variations in oxygen isotope ratios ($^{18}O/^{16}O$) of the deep-sea sediment cores noted above, but they contained the same periodicities as the Milankovitch cycles. Subsequent ice cores from Greenland confirmed these same patterns. A remarkable fact had now emerged—the ice from both the polar ice caps and sediments from the deep oceans all recorded past climate change, and they all showed the same underlying periodic patterns predicted by Milankovitch.[3,5,6] The results from the study of deep-sea cores and continental ice cores surely represents one of the triumphs of modern science in the late twentieth century.

One such ice core is called the Greenland Ice Sheet Project 2 (GISP2).[7] It is of particular interest to us here because it spans the four cultural periods discussed in chapter 2. It was drilled near the center of the Greenland ice sheet, and after five years of drilling, 3,000 meters (9842 feet) of core was retrieved in 1993. The winter temperature was about -20°C (-4°F) at the drilling site, indicating how difficult the working conditions were, while at the same time, workers avoided contamination or damage to the core, making its retrieval a remarkable engineering and scientific feat. The core is now stored at the National Ice Core Laboratory in Denver. The entire core goes back 110,000 years, which covers the most recent glacial cycle (out of a total of eight such cycles) of the last Ice Age. The individual annual layers were counted back to about 100,000 years, so this is one of the most accurate timescales available for this time period.

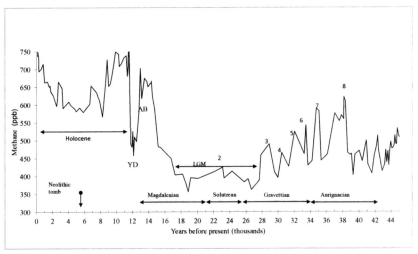

Figure 3.1: Graph of methane concentration in the GISP2 (Greenland Ice Sheet Project). Concentrations are in parts per billion. The four Upper Paleolithic cultural periods are indicated. The length of these periods varies from place to place. The numbered peaks refer to Greenland interstadials. YD–Younger Dryas; AB–Allerød-Bølling interstadials; LGM–Last Glacial Maximum.

Figure 3.1 shows the variation of methane in the GISP2 ice core back to 45,000 years ago. Carbon dioxide (and also oxygen isotope ratios) show very similar patterns to the methane gas pattern, indicating that these two gases faithfully tracked climate change during the last Ice Age. Archeologists commonly use the oxygen isotope ratio in ice cores or deep sea sediment cores to calibrate their timescales.[8,9] Methane is chosen here over oxygen isotopes because the causes of the methane variations in the global cycle are related to continental climate conditions, whereas oxygen isotopes ratios are related more to ocean dynamics, which are more complex and less well understood. In addition, when the methane record is combined with atmospheric dust concentrations (indicating aridity), a more detailed picture of Ice Age climate conditions can be reconstructed (see below). Studies of tree pollen in deep-sea sediment cores in the Mediterranean reveal very similar climate patterns to those in the GISP2 ice core, indicating that the Greenland core reflects climate changes over much of Europe during the last Ice Age.[10]

Before proceeding, a note on terminology is in order. Major Ice Age cold and warm cycles are referred to as glacial and interglacial periods. But these glacial cycles are punctuated by shorter millennial-scale cycles which are referred to as stadials (cold periods) and interstadials (warmer periods), to distinguish them from the longer glacial-interglacial cycles. The overall pattern in the ice cores is one of prolonged cooling cycles followed by abrupt warming, to give a sawtooth shape to the curves. The sawtooth pattern is especially noticeable in deeper cores, such as the Vostock core.[5] The warm periods (interstadials) are numbered two through eight in figure 3.1, and these interstadials correspond to "teeth" in the saw.

In figure 3.1, high methane concentrations (about 750 ppb) correspond to interstadial conditions (warm periods) and low methane concentrations (about 350 ppb) correspond to stadials (cold periods). For example, during the Allerød-Bølling warming (AB, figure 3.1) at the end of the Last Glacial Maximum, which began about 14,700 years ago, January temperatures increased by about 20°C (36°F) from about minus 20°C (-68°F) to 0°C (32°F).[11] The change in atmospheric methane was from about 400 ppb to 650 ppb. At the end of the Younger Dryas (YD, figure 3.1) at 11,500 years ago the change in temperature was also very abrupt and occurred on a time scale of thirty to forty years[12], equivalent to the lifespan of an Upper Paleolithic adult. The oscillations in methane concentrations prior to the Younger Dryas indicate a period of very unstable climate corresponding to the Magdalenian, Solutrean, Gravettian, and Aurignacian cultural periods and may have occurred on a similar time scale as the Younger Dryas, although the time resolution is not as good as the ice gets older. These rapid oscillations lasting a few thousand years are too fast to be accounted for by Milankovitch cycles, which occur over periods of tens of thousands of years (see bullets above).

By studying deep-sea sediment cores in the North Atlantic, scientists attribute the millennial-scale climate oscillations to calving of large ice blocks off the North American ice sheet into the North Atlantic during cold periods during the last Ice Age. Melting of these rafted blocks produced fresh melt water, resulting in lower salinity of the seawater (which is normally about 3.5 weight percent) and caused the Gulf Stream current, which brings heat from the tropics to northern latitudes in the Atlantic, to come to a standstill. This resulted in rapid cooling of the North Atlantic and also of northern Europe. Following these cold conditions, abrupt warming appears to have occurred, causing the ice to retreat and the Gulf Stream to restart, returning warm water to higher latitudes again.

Deep-sea sediment cores normally consist of layers of very fine-grained sediment, such as silt and clays. Several cores from the North Atlantic at a latitude of 45–50°N, however, show several distinct layers of coarse-grained sand that could only have come from the continents. These layers, when dated, correspond to the cold periods in the ice cores (such as GISP2), and it was concluded that the only way they could have been deposited was from floating icebergs that broke off the continental ice sheet in Canada and rafted into the North Atlantic Ocean during cold periods. They then melted, to drop the continentally derived sediment embedded in the underside of the floating ice. These ice-rafted sediments correspond to the very cold periods in the Greenland ice core and are referred to as Heinrich events in the literature. This research, combining the study of continental ice cores and deep-sea sediment cores, provides clear evidence that the atmosphere, the oceans, and the ice sheets together act as a closely linked, complex system, but the causes and effects between ice sheets, greenhouse gases, and ocean currents are not yet fully understood.[13] Detailed studies of the ice cores indicate that the greenhouse gases sometimes increase *after* the temperature changes, indicating that these gases were not the primary cause of the changes but they subsequently amplified those changes.[12]

When Greenland ice cores are combined with precise carbon-14 dates, events in the archaeological record can be identified with specific climate events as far back as the Aurignacian. From figure 3.1, it is apparent that during the early Aurignacian prior to interstadial number eight, there were several stadials (cold cycles) on a millennial scale in the interval 40,000 to 42,000 years ago. An important Aurignacian archaeological site in the Dordogne region of southwest France is the rock shelter Abri Pataud. Here, on the basis of sediment, tree pollen, and studies of animal bones a cold-dry period was identified.[14] Precision dating of this level yields an age range of 41,000 to 42,000, corresponding to a stadial (cold period) on the Greenland ice core.[9] Without accurate radiocarbon dates that are calibrated together with chronologies provided by ice cores, such reconstructions were not possible until quite recently. It is easy to see how an error of ± 500 years in timing can cause a mismatch between cold and warm periods. This topic is taken up again in chapter 7.

What were the causes of the variations in atmospheric methane during the last Ice Age and what can they tell us about the environmental conditions at that time? Since industrial times (1860s), the methane concentration in the Earth's atmosphere has increased almost three times over its pre-industrial value. The cause of this increase since the nineteenth century is due mainly to industrialization and more intensive agriculture worldwide. Major manmade sources of methane today include energy extraction (coal mining, oil and gas production), landfill waste, farm animals, rice paddies, and forest fires. Rice paddies are wetlands that make up a substantial fraction of agricultural land in Asia.

The main natural source of methane prior to the Industrial Revolution was wetlands, both tropical (for example, the Florida Everglades and Indonesian coastal swamps today) and northern hemisphere sources, such as Canadian and Siberian peat bogs. It is commonly assumed that similar continental sources produced the variations of methane seen in the GISP2 ice core.[15,16]

Natural wetlands, the chief cause of pre-industrial methane emissions, go by a variety of names, including mires, bogs, peat lands, swamps, marshes, and fens, but these fine distinctions need not concern us here. In wetlands, bacteria break down complex organic molecules to produce simpler molecules, such as acetic acid (or vinegar), under oxygen-poor conditions. The acetic acid (CH_4CO_2) in turn is broken down to produce two greenhouse gases: methane (CH_4) and carbon dioxide (CO_2).

As with all chemical reactions, higher temperatures promote faster reaction rates. Higher temperatures during wet periods would therefore be expected to produce high concentrations of methane in the atmosphere, during monsoon rains, for example. Once the wetlands go dry, however, these reactions are shut down, and methane is no longer produced. Dry and cold conditions would therefore produce low methane concentrations in the atmosphere. The methane curve in the GISP2 core can provide evidence of the type of climate in the past—high concentrations of methane pointing toward warm moist conditions, and low methane concentrations pointing

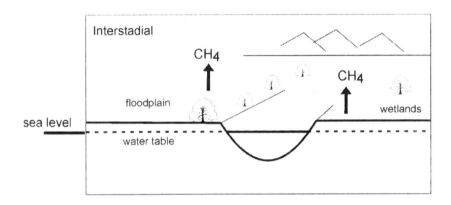

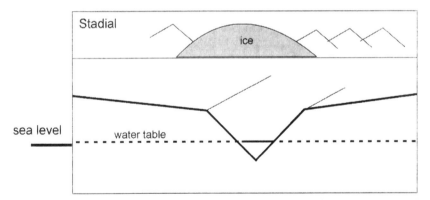

Figure 3.2 Upper panel: During interstadial periods, the climate is moist and temperate. Sea level is similar to today and the water table is high. Wetlands (floodplains, peat bogs, swamps) contribute methane to the atmosphere. **Lower panel:** During stadial periods, sea level is lower due to ice on the continents. The water table is lower and wetlands are dry, and atmosphere methane levels are low. Rivers have a V-shape profile, indicating downward erosion.

toward cold and dry conditions (see figure 3.1). Because the air carries less moisture when it is cold, cold tends to go hand-in-hand with dry conditions, while warm air tends to go with moist conditions.

During the Last Glacial Maximum (LGM, figure 3.1), the ice sheets extended as far south as the fortieth parallel in North America (just south of Cincinnati in the Midwest) and to the fiftieth parallel in Europe, corresponding to the southern tip of England and the northern parts of continental Europe. The ice sheets were about 3 km (~2 mi) thick, and this volume of water was transferred out of the oceans and deposited on the continents. As a result, sea level was about 100 meters (330 feet) lower during the LGM than it is today.

Coastlines were farther out than their present location, and rivers carved canyons into the continental shelf. Ireland was connected to England, and England was connected to the continent at this time; in North America, a land bridge existed across the Bering strait, thought by many to have been a pathway for modern man to colonize North America from Asia. On land, rivers eroded downward (indicated by their V-shaped cross-section) until they reached the new, lower sea level at that time. Because drainage levels were lower, wetlands dried up. This would have cut off the supply of methane to the atmosphere. During times of glacial melting, sea level rose again, the climate became wetter due to more active tropical monsoon rains, and wetlands began to produce methane again. Methane levels therefore closely track climate changes during the glacial cycles. These two different situations (high sea level and low sea level) are summarized in figure 3.2.

The GISP2 ice core provides us with additional details on the climate during this time period. Figure 3.3 is the same as figure 3.1, with the addition of dust levels in the ice core shown on the right-hand scale.[7,17] A remarkable pattern is apparent—dust levels are high when methane levels are low, and conversely, dust is low when methane levels are high. These data confirm the scenario outlined above. During cold periods (low methane emissions), ice accumulated in the polar regions, sea level dropped, wetlands became dry, and wetland vegetation died out. Wind then eroded the exposed soils and transported the dust particles into the atmosphere. This dust was deposited as sedimentary layers called *loess* (pronounced luss) by geologists. The dust also reached Greenland and was deposited in the ice. Conversely, during warm periods (high methane levels), polar ice melted, sea level rose, wetlands became saturated with water and plant life returned, thereby stabilizing the soils with their root systems, and atmospheric dust was reduced.

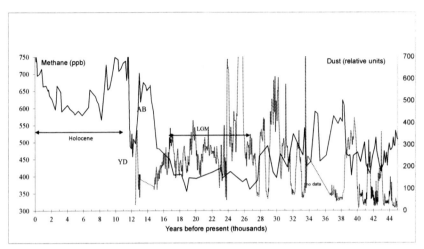

Figure 3.3: The same as figure 3.1, with the addition of dust levels (right-hand scale) in the GISP2 ice core. High dust levels correspond to low methane periods (stadials) indicating a cold, dry climate. Low dust levels correspond to high methane levels, indicating a temperate, moist climate.

In summary, figure 3.3 indicates that during interstadial periods, the climate was temperate and moist, whereas during stadials, the climate was cold and dry with a lot of dust in the atmosphere. Studies of sediments and tree pollen and animal remains from Aurignacian and Magdalenian sites in France confirm these climate swings.[18] Also, a study of pollen from a marine core from the western Mediterranean shows oscillations between temperate warm-moist climate and cold semi-desert climate.[10] Dust storms must have been a serious impediment to the Upper Paleolithic hunters of the time, and, together with the cold, life on the tundra-steppe would have been harsh if not impossible. Paleolithic hunters retreated to less hostile refuges farther south (northern Spain and southwest France) during stadials.[19]

Based on studies of ocean cores and ice cores, scientists now understand reasonably well the glacial environments and the climate zones that existed during the Upper Paleolithic in Europe from 45,000 to 10,000 years ago. This puts us in a good position to examine the habitats of the large animals that dominate Franco-Cantabrian cave art in the next chapter.

References cited

1. Imbrie, J. and Imbrie, K. P., 1986. *Ice Ages, Solving the mystery.* Harvard University Press.
2. Drake C. L., Imbrie, J., Knauss, J. A., and Turekian, K. K., 1978. *Oceanography.* Holt, Rinehart & Winston.
3. Hays, J. D., Imbrie, J. and Shackleton, N. J., 1976. "Variations in the Earth's orbit: pacemaker of the Ice Ages." *Science*, 194, 1121–1132.
4. Stocker et al., 2013 "Climate Change: The Physical Science Basis." *Intergovernmental Panel on Climate Change*. Cambridge University Press.
5. Pettit, J. R. et al. 1999. "Climate and atmospheric history of the past 420,000 years from the Vostok ice core, Antarctica." *Nature*, 399, 429–436.
6. Bond G. et al. 1993. "Correlations between climate records from North Atlantic sediments and Greenland ice." *Nature*, 365, 143–147.
7. *The Greenland summit ice cores* CD-ROM. 2003. National Snow and Ice Data Center and NOAA, Boulder, CO.
8. Banks, W. E., D'Errico, F. and Zilhão, J. 1993. "Human-climate interaction during the early Upper Paleolithic: Testing the hypothesis of an adaptive shift between the proto-Aurignacian and the Early Aurignacian." *J. Human Evolution*, 64, 39–55.
9. Higham, T. et al. 2011. "Precision dating of the Paleolithic: A new radiocarbon chronology for the Abri Pataud (France), a key Aurignacian sequence." *J. Human Evolution*, 61, 549–563.
10. Nebout, C. N. et al. 2002. "Enhanced aridity and atmospheric high-pressure stability over the western Mediterranean during the North Atlantic cold events of the past 50 ky." *Geology*, 30, 863–866.
11. Renssen, H. and Isarin, R. F. B., 2001. "The two major warming phases of the last deglaciation at 14.7 and 11.5 ka cal BP in Europe: climate reconstructions and AGCM experiments." *Global and Planetary Change*, 30, 117–153.
12. Severinghaus, J. P. et al. 1998. "Timing of abrupt climate change at the end of the Younger Dryas interval from thermally fractionated gases in polar ice." *Nature*, 391, 141–146.
13. Bond, G. C. and Lotti, R. 1995."Iceberg discharges into North Atlantic on millennial time scales during the last glaciation." *Science*, 267, 1005–1009.
14. Blades, B. S., 2001. *Aurignacian Lithic Economy. Ecological perspectives from southwestern France.* Kluwer/Plenum, New York.
15. Brook, E. J. et al. 2000. "On the origin and timing of rapid changes in atmospheric methane during the last glacial period." *Global Biogeochemical Cycles*, 14, 559–572.

16. Chappellaz, J. et al. 1993. "Synchronous changes in atmospheric CH_4 and Greenland climate between 40 and 8 kyr BP." *Nature*, 366, 443–445.
17. Ram, M. and Koenig, G. 1997. "Continuous dust concentration profile of pre-Holocene ice from the Greenland Ice Sheet Project 2 ice core: Dust stadials, interstadials and the Eemian." *J. Geophysical Research*, 102, 26, 641–26, 648.
18. Leroi-Gourhan, Arl., 1997. "Chauds et froids de 60,000 a 15,000 BP." *Bull. Soc. Préhistorique Francaise*, 94, 151–160.
19. Bocquet-Appel, J-P., 2000. "Population kinetics in the Upper Paleolithic in Western Europe." *J. Archaeological Science*, 27, 551–570.

4 Ice-Age Herbivores

Since all living things require a certain minimum amount of warmth and water, resources that are largely controlled by climate, climate primarily determines plant and animal geography. Vegetation types change according to climate zones, which in turn are largely determined by latitude, altitude, and moisture. Alexander von Humboldt was a German naturalist who, in his botanical studies in the Andes in the eighteenth century, was one of the first scientists to document the effect of elevation on different plants.[1] Animals migrate according to seasonal changes and will also adapt to longer-term climate change. They follow food sources as vegetation zones change and also migrate to seek more favorable temperature conditions.

Today's global warming gives us a preview of some of the changes in the distribution of plants and animals that occurred during the last Ice Age, but in the opposite direction.[2] In Europe, New Zealand, and Alaska the tree line has moved toward higher altitudes due to warming. Arctic shrub vegetation in Alaska has also shifted to higher elevations. Plant diversity has increased substantially over the past hundred years on Swiss peaks in the Alps, due to the migration of plant species to higher elevations. In Canada, the red fox has expanded its range northward, but the arctic fox's range has retreated. The arctic fox was common in the south of France during the last Ice Age, as indicated by bone remains at the La Vache cave, north of the Pyrenees. Less sea ice in the arctic today has increased coastal erosion, forcing entire towns to move back from the coastline, an example of human migration due to climate change. As pointed out in the previous chapter, human populations moved south to refuges during the Last Glacial Maximum.

Computer models of forests indicate there is a time lag of about 150 to 200 years before forests reestablish themselves after a rapid swing in the climate.[3] In general, during a warming period, forests can be expected to migrate north or to higher elevations to offset the change. Conversely, during colder periods forests migrate south or to lower elevations. The rate at which these migrations take place will likely vary from species to species, and it depends on their tolerance to unfavorable conditions. In the case of animal migrations, this can lead to mixed faunal assemblages in times of rapid climate change. By today's standards, meaning the Holocene epoch, the Ice Age bestiary was complex, involving a large assortment of animals we would not normally expect to associate together, such as: woolly mammoths, woolly rhinoceros, red deer, reindeer, wild cattle (aurochs), horse, bison, lions, and bears. Nothing comes close to this bizarre assemblage of animals today, including the fauna of Alaska and Siberia.[4]

Because there is no equivalent environment today for some of the climate zones that existed during the last Ice Age, researchers created two new climate categories for the Last Glacial Maximum: A northern tundra-steppe and a southern tundra-steppe. The northern tundra-steppe was a polar desert, and it existed immediately south of the continental ice sheets.[5] The southern tundra-steppe, also called the Mammoth Steppe, extended all the way from Western Europe east through central Asia to Siberia and on to Alaska, ranging from 40° N along its southern margins north to 70° N along its northern margin.[6] Steppe refers to flat, grassy but treeless plains, and tundra refers to a region of permanently frozen soil (permafrost). The growing season was short, and cold, dry winds swept the landscape. There was little vegetation apart from grasses, sedges, lichens, and mosses. On the southern tundra-steppe, the maximum temperatures were below 10°C (50°F), with some melting in the warmer months. By contrast, the present-day conditions in these mid-latitude regions are described as temperate broadleaf forest (prior to modern farming activities). This would correspond to Central Europe, including central and southern France. Further south, in Spain for example, the climate was semi-desert and dry steppe. Further south again, fully fledged desert dominated, such as the Sahara. During temperate periods (interstadials), a broadleaf forest moved north to replace the southern tundra steppe, and more temperate animals dominated the fauna (aurochs, red deer, and ibex) together with the horse, which was apparently cold-adapted but also tolerated temperate climates (see chapter 5).

Despite the sparse vegetation, the southern tundra-steppe supported three large herbivores, the woolly mammoth (*Mammuthus primigenius*), the bison (*Bison pricus*), and the horse (*Equus caballus*). These three animals were all grazers (mainly grass diet) rather than browsers (bark and branches). Intact specimens of woolly mammoth, bison, and horse of Paleolithic age have been recovered from frozen northern tundra environments, such as Siberia and Alaska, some with food remains in their stomachs—mainly grasses.[6] Of these three large herbivore species, both the woolly mammoth and the horse became extinct at the end of the Pleistocene, but the bison survived into the Holocene, evolving into the modern European and American bison. A central-Asian horse that lives on the Mongolian steppe today is thought to be the closest living relative of the extinct Ice Age horse. These extinctions occurred at the Holocene-Pleistocene transition, about 12,000 years ago, and much of the debate over the extinctions centers on whether they were caused by excessive hunting by man, climate change, or both.[7] The big-game hunter and zoologist R. D. Guthrie, notes that the horse, rhino, and mammoth have similar digestive systems that allow a rapid throughput of poor-quality forage, which gave them an advantage on the Mammoth Steppe. The warmer and wetter conditions of the Holocene epoch promoted more toxic plants, and these animals were at a disadvantage compared to bison, elk, and moose, which flourished after the Holocene transition, but not before.[6,7]

Animal species commonly depicted in Upper Paleolithic cave art include the bison (*Bison priscus*), the horse (*Equus caballus*), the mountain goat (*Capra ibex*), the red deer (*Cervus elaphus*), the aurochs (*Bos primigenius*), the woolly mammoth (*Mammuthus primigenius*), the reindeer (*Rangifer tarandus*), the woolly rhinoceros (*Coelondonta antiquitatis*), and the giant elk (*Megaloceros giganteus*); carnivores such as lions and bears are among the least-common animals depicted. These animals are described in more detail below.

Woolly Mammoth

The Pleistocene woolly mammoth (*Mammuthus primigenius*) evolved from an ancestor that was acclimatized to a warmer woodland habitat.[8] The newly evolved woolly mammoth was well adapted to the rigors of glacial episodes with a thick layer of fat under its skin and two coats—a short undercoat and a longer, shaggy overcoat that hung down to the ground. It had smaller ears than an African elephant and a shorter tail; it also had an anal flap to conserve internal heat; this flap is visible on a mammoth depicted at Rouffignac.[8] Analysis of fossil teeth of woolly mammoths show traces mainly of grasses, and their stomach contents also indicate a grass diet. One of the best-preserved woolly mammoths was found frozen in the tundra of Siberia in 1977; it is that of Dima, a baby mammoth.[6] Pollen surrounding this find indicated an herbaceous, dry tundra-steppe environment, consistent with mammoths as grazers, rather than browsers like moose, for example. Dima is thought to have fallen into a muddy pond during a summer thaw on the tundra and drowned. Apart from humans, woolly mammoths probably had few predators, with the possible exception of lions. At the Baume Latrone cave in the department of Gard, in southern France, a lion and a mammoth are depicted on the same panel, a rare example of this particular prey-predator pairing.[9] In the Pleistocene of Central and Eastern Europe, mammoths were an important resource on the treeless tundra-steppe, where wood was in short supply and mammoth bones were used to construct shelters. These bones were from hunted or trapped animals or from those that died a natural death.

The fossil record of mammoth-bone sites in Europe is extensive, with at least 750 locations spanning the interval 50,000 to 13,000 years ago, after which time they became extinct.[10] Their extinction in North America is dated to about 11.5 ka ^{14}C BP (13.3 ka cal BP)[7], and in Europe to about 12.0 ka ^{14}C (14.2 ka cal BP). This time period corresponds to the Allerød-Bølling warming (AB in figure 3.1). In Europe, the last mammoths appear to be those engraved on slate at the Magdalenian open air site of Gönnersdorf in Germany, where most of the mammoths, although otherwise well-illustrated, are without tusks.[10] The zoologist R. D. Guthrie attributed this to poor nutrition before their extinction. The mammoth did not adapt fast enough to the rapid change in new vegetation, which defended itself against herbivores by becoming toxic, at the end of the last glacial.[7] This appears to favor attributing extinction to climate change, rather than overhunting by man. A global study of mammoths over a longer timeframe, including earlier mammoth species, concludes, however, that human overkill was responsible for several mammoth species extinctions.[11]

During its peak, the global range of the woolly mammoth was immense, spanning 270 degrees of longitude and 45 degrees of latitude, including all of Europe, North America, and Asia.[10] In the west, they reached the British Isles and Scandinavia, and in the east, Eastern Europe, the Urals, Siberia, and the Bering Strait; to the south they reached southern Spain (37°N latitude) and central Italy. A synthesis of mammoth migration patterns based on 380 carbon-14 dates in Europe and Asia shows the following broad patterns over the past 50,000 years.[12] In the interval 50–46 ka cal BP, most bone locations are concentrated in Central Europe. During 46–43 ka cal BP, they migrated north into the British Isles when sea level was lower during cold phases. During warmer periods in the interval 43–41 ka cal BP, they expanded into ice-free areas covering a larger range. When Scandinavian ice sheets expanded south, mammoths migrated south into northern Spain, in the interval 36-33 ka cal BP.

In southern Spain, peat bog deposits have preserved coexisting bones of mammoth, bison, and red deer, yielding ages in the range 40–30 ka cal BP.[13] Mammoth and red deer are an example of a mixed fauna—the mammoth being an indicator of cold climate and the deer an indicator of a more temperate climate. Paleobiologists have noted that, during the extreme climate conditions of the last glacial, faunal assemblages became mixed. At least four Upper Paleolithic caves depict this type of anomalous mixed faunal assemblage, with at least one mammoth depicted with red deer in the same cave: These include Cougnac (Lot, France), Ebbou (Ardèche, France), and Altamira and El Castillo (both in Santander, Spain).

A second, much later expansion of the mammoth into Spain occurred during the Last Glacial Maximum about 24,000 to 19,000 years ago. After 24,000 years, there is a gap in carbon-14 dates of mammoth bones for this interval in Western Europe.[14] The migration route south appears to have been around the Pyrenees along the Bay of Biscay on the Atlantic side, and an eastern route on the Mediterranean side, used when sea levels were lower. Apparently, conditions were too inhospitable even for the woolly mammoth farther north in Western Europe at that time. This time period includes much of the Solutrean and the early Magdalenian cultures. The mammoth-bone record reappears again at about 19 ka cal BP until the mammoth's final extinction, noted above at about 14.0 ka cal BP. Only five locations of mammoth bones are known in Europe from after the Last Glacial Maximum.[12] Mammoth migrations are taken up again in chapter 9.

Woolly Rhinoceros

Now extinct, the woolly rhinoceros (*Coelodonta antiquitatis*), a cold-adapted animal, had a thick, woolly fleece and, like the mammoth, was also an herbivore, eating mainly grasses on the tundra-steppe. It had two unequal horns, one in front of the other. They are rarely depicted in cave art; the exception being the Chauvet cave, where the rhino outnumbers (sixty-six depictions) the woolly mammoth (sixty-eight). One of the rhinos at Chauvet yields a direct radiocarbon age of 31 ka ^{14}C BP years (or about 35 ka cal BP).[15] They had few natural predators and were probably not hunted much because of the danger involved.

The Bison

Bison pricus reached two meters (6 feet 6 inches) in height, had a hump over its shoulders, and had long horns and a powerful jaw. Based on frozen mummies from Alaska, we know that the Pleistocene steppe bison had a different appearance than the American (*Bison bis*on) or European bison (*Bison bonasus*), although they may represent a single complex species.[6] Different types are depicted in cave art, some with shaggy coats, possibly reflecting different climate conditions, seasons, or different artistic styles. At Lascaux, in the shaft scene, a wounded bison appears to be charging a male hunter. Bison are very well cold-adapted and even have hair on their lower limbs; they have short tails and furry ears and a thick, dark winter coat, which molted in the spring to a lighter color. They are grazers and prefer grass but also eat branches, bark, lichen, moss, and acorns (browsing). They also lived on the tundra-steppe. Unlike caribou (or moose) who are browsers, bison have difficulty feeding in deep snow because of their shorter legs. However, this may not have been a problem for them, as the glacial winters were cold and dry, without much snow on the ground.

Genetic studies of the ancient DNA of steppe bison together with ^{14}C dates on bones show that their population numbers increased rapidly over the period 75 ka cal BP to 37 ka cal BP, followed by a slow decline beginning at about 37 ka cal BP, corresponding roughly to interstadial eight (see figure 3.1), when warmer conditions prevailed, reduction in tundra-steppe occurred, and tree cover returned.[16] Bison are not well-suited to woodlands, which provide no food sources for them.[6]

The Horse

The Przewalski or tarpan horse (a central-Asian wild horse that lives on the Mongolian steppe) is thought to be the closest surviving relative to the extinct ice-age horse (*Equus caballus*). They also lived in parts of Europe during the Middle Ages. They were broad-backed and had short legs with a fat, low belly; those depicted at Lascaux are referred to as Chinese horses. They developed two coats per year, including a shaggy winter coat, and were suited to the cold, although possibly not as cold-adapted as the woolly mammoth or woolly rhinoceros. A mummified Pleistocene horse was found in Siberia in 1968 and is thought to be very similar to the Przewalski horse. It yielded a radiocarbon age of 35-39 ka ^{14}C BP. It had a tail with long hair and a reddish-brown coat. It died with a full stomach, consisting mainly of grass, indicating that it was a grazer rather than a browser. These horses are known to be aggressive, fast, and difficult to hunt. As noted in chapter 2, they were hunted year-round in some Magdalenian encampments. The horse is the most commonly depicted animal in Upper Paleolithic cave imagery.

Aurochs

The aurochs (*Bos primigenius*), a type of wild cattle or bovine, was an ancestor of the domestic oxen, and modern domesticated cattle also appear to be descended from aurochs. Prehistoric varieties were over 1.6 meters (5 feet) tall and weighted up to a 1,000 kg (2,220 lbs.) and had large, sharp horns. First noted by Julius Caesar in his account of the Gallic War, it was described as very strong, fast, and dangerous. They were also noted in the wild in the Pyrenees. Like modern cattle, they were mainly grazers but not as cold-adapted as bison or the woolly mammoth. They would have lived on open woodlands and grasslands, corresponding to the more temperate climate south of the Mammoth Steppe.

Red Deer

Red deer (*Cervus elaphus*) are quite common in cave art and far outnumber reindeer. Of the red deer, only males have antlers, and they shed them once a year; in the case of reindeer, both sexes have antlers. In cave art depictions, the deer is one of the few animals that can be identified by gender. They are commonly depicted in northern Spanish caves, possibly because of the milder climate. Red deer prefer open wooded grassland, similar to that favored by the aurochs.

Reindeer

The reindeer (*Rangifer tarandus*) has backward-pointing antlers, which distinguish it from the red deer stag. Although reindeer are well-represented in kitchen waste piles of excavated archaeological sites, indicating they were an important part of the hunters' diet, particularly in the Magdalenian period, they are not commonly depicted in cave art. Henri Breuil, who favored the hunting magic theory for cave art, suggested that, because reindeer were easier to hunt, compared with the red deer, for example, there was no need to portray them in cave art, even though they were an important part of the diet. At the Magdalenian reindeer hunting camp at Pincevent near Paris, described in chapter 2, the reindeer were thought to have been hunted as they crossed the River Seine on their seasonal migration route. Reindeer today are restricted to the tundra-steppe of Siberia and Scandinavia, similar to caribou of Canada and Alaska, and are cold-adapted.

The giant deer (*Megaloceros giganteus*) stands about 2 meters shoulder high (6½ feet), and its antlers span up to 3.6 meters (about 12 feet). Also called the Irish elk, a full skeleton greets the visitor in the foyer of the National Museum of Ireland in Dublin. It was common in Europe as far east as Lake Baikal at the end of the last Ice Age but became extinct in most of Europe during the exceptionally cold period of the Younger Dryas, 12,000 years ago.[14] It survived in Siberia up until about 7,000 years ago, in the Holocene. It was a mixed feeder, requiring both grazing and browsing. Its giant antlers would make closed forest an unsuitable habitat. The fossil record of the giant deer (*Megaloceros*) shows an even longer time gap centered on the Last Glacial Maximum, when compared to that of the mammoth, outlined above,[14] suggesting it was less well cold-adapted than the mammoth. With exception of the Cougnac cave in Lot, France, where three are portrayed, it is relatively rare in cave art.

Ibex

The ibex (*Capra ibex*) or mountain goat is fairly common in European cave art. They are great climbers and prefer high altitudes. There were two types of ibex, an Alpine type (*Capra ibex*) and the Pyrenean species (*Capra pyrenaica*), which have horns with different cross-sections. During cold periods, when glaciers in the Alps and the Pyrenees expanded, they would have been forced onto lower ground where they were hunted. Ibex are quite common in northern Spain cave art, reflecting the proximity of the Pyrenees, and ibex hunting camps have been identified in Solutrean deposits of northern Spain.[17]

A synthesis of studies of the remains of hoofed animals, including red deer, steppe bison, ibex, and the horse, in the Aquitaine region in southwest France during the latter part of the Last Glacial Maximum (LGM) shows the important role of terrain in determining the distribution of these herbivores, in addition to climate.[18] The Aquitaine region is bordered to the east by the Atlantic Ocean and to the west by the Massif Central, which approaches 1.8 km (6,000 feet) in elevation, and it is bounded to the south by the Pyrenees (~ 3 km elevation). This large area includes some of the most important cave art sites in all of France, including Lascaux, Rouffignac, and Font de Gaume. The cities of Bordeaux mark the northwest margins of the area, and Toulouse lies to the southeast (see location map at the end of the book).

The climate in this region at the time of the LGM was similar to that of North Dakota today—an average summer maximum temperatures of 15°C (59°F) and an average minimum winter temperature of minus 15°C (5°F). The topography of this region can be divided into three areas:

- Zone 1: A coastal zone extending inland with an elevation of 0–200 meters (650 feet),
- Zone 2: A higher elevation region to the west, toward the foothills of the Massif Central, 200–500 meters (650–1,600 feet), and
- Zone 3: 500 meters-plus (1,600 feet-plus), including the Pyrenees and the Massif Central

The density distribution of reindeer bones in this region shows a concentration centered over zone two to the north and a decrease to the west toward Bordeaux—apparently, reindeer in this environment favored higher altitudes away from the coast. The reverse pattern is shown by the bison—the maximum concentration is centered over the Bordeaux region and decreases to the east toward the Massif Central, so it appears that bison favored low-relief terrain. The horse shows a very similar distribution to that of the bison. As might be expected, the ibex showed preference for higher elevation toward the Massif Central and the Pyrenees to the south. These studies indicate that bison and horse favored lowland areas, whereas the reindeer tolerated higher altitudes, and the ibex favored high relief, rocky slopes, and mountainous regions. A more recent study depicts the Upper and Middle Magdalenian (time frames somewhat later than the Last Glacial Maximum) bone assemblages for southwest France, and they show a broadly similar pattern to that outlined above.[19]

References cited

1. Jackson, S. T., 2009. "Alexander von Humboldt and the general physics of the Earth." *Science*, 324, 596–597.
2. Walther et al., 2002. "Ecological responses to recent climate change." *Nature*, 416, 389–395.
3. Davis, M. B. and Botkin, D. B., 1985. "Sensitivity of cool-temperate forests and their fossil pollen record to rapid temperature change." *Quaternary Research*, 23, 327–340.
4. Butzer, K. W., 1964. *Environment and Archaeology*. Aldine publishing, Chicago.
5. Adams, J. M. et al. 1990. "Increase in Terrestrial Carbon Storage from the Last Glacial Maximum to the Present." *Nature*, 348, 711–714.
6. Guthrie, D., 1990. *Frozen Fauna of the Mammoth Steppe*. Chicago University Press.
7. Guthrie, D., 2006. "New carbon dates link climate change with human colonization and Pleistocene extinction." *Nature*, 441, 207–209.
8. Lister, A. and Bahn, P., 1994. *Mammoths*. Macmillan, New York.
9. Leroi-Gourhan, A., 1967. *Treasures of Prehistoric Art*. Abrams, Inc., New York.
10. Braun, I. M. and Palombo, M. R., 2012. "*Mammuthus primigenius* in the cave and portable art: An overview with a short account of the elephant fossil record in southeastern Europe during the last glacial." *Quaternary International*, 276–277, 61–76.
11. Surovell, T. et al., 2005. "Global Archaeological Evidence for Proboscidean Overkill." Proceedings National Academy Sciences, 102, 6231-6236.
12. Markova, A. K. et al., 2010. "New data on the dynamics of the *Mammuthus primigenius* distribution in Europe in the second half of the late Pleistocene-Holocene." *Doklady Akademii Nauk*, 431, 547–550.

13. Alvarez-Lao, D. J. et al., 2009. "The Padul mammoth finds—on the southernmost record of *Mammathus primgenius* in Europe and its southern spread during the late Pleistocene." *Paleoclimatology*, 278, 57–70.
14. Stuart, A. J. et al., 2004. "Pleistocene to Holocene extinction dynamics in giant deer and woolly mammoth." *Nature*, 431, 684–689.
15. Valladas et al., 2005. "Bilan des datations carbone 14 effectuees sur les charbons de bois de la Grotte Chauvet." *Bull. Soc. Préhistorique Francaise*, 102, 109–113.
16. Shapiro, B. et al., 2004. "Rise and fall of the Beringian steppe bison." *Science*, 306, 1561–1564.
17. Straus, L. G., 1990. *The glacial maximum in Cantabrian Spain: The Solutrean in The World at 18,000 BP*. Soffer, O. and Gamble, C., eds., Unwin Hymen, London.
18. Delpech, F. 1990. "The range distributions of some ungulate mammals during the Dryas I (recent Wurm) in Aquitaine: Paleoecological implications." *Geobios*, 23, 221–230.
19. Langlais, M. et al., 2012. "The evolution of Magdalenian societies in southwest France between 18,000 and 14,000 cal BP: Changing environments, changing tool kits." *Quaternary International*, 272–273, 138–149.

5. Cold- and Warm-Adapted Animal Groups

A statistical technique called principal component analysis (or factor analysis) has long been used in educational literature, where large numbers of variables are often involved. In this method, the variation between the variables is explained by reducing the dimensions of the problem to a few variables, without losing too much of the information in the data.[1] If the variation in the data can be explained by two or three variables, called principal components, then the results can be conveniently displayed on two-dimensional plots. In our case, we want to see if the variation in the distribution of the nine herbivores commonly depicted in cave art (reviewed in the previous chapter) can be explained by a few underlying, controlling variables. The results of such an analysis only indicate the amount of variation that the principal components can explain, and they do not provide any physical insight into what those variables might be—it is up to the interpreter of the data to make that decision. Obviously, the interpretation should make physical sense.

The first step in the statistical procedure is to calculate the variance in individual animal numbers depicted in as many caves as possible, in our case thirty-two caves. The variance is the square of the difference between the mean number for a given animal depicted in many caves and the observed number in any given cave. For example, if the horse occurs on average eight times in all caves but occurs twelve times in a specific cave, the variance for that cave is eight minus twelve, squared, or sixteen. Next, a best-fit line is found that explains the largest amount of the variance in all the data using linear regression. This line becomes the first principal component. The next step is to find another line at right angles (or orthogonal) to the first line that explains the next-largest part of the variance. Mathematically, this component is independent of the first component because it is orthogonal, and it is labeled the second principal component. And so on, with a third line, again orthogonal to the other lines, until we can explain a large portion or all of the variance in our data. In two dimensions, these principal component axes are usually rotated along with the data to correspond to the X and Y axes on a graph, thereby making interpretation visually simple. Humans are not very well-equipped to deal with these mathematical gymnastics, especially beyond two dimensions, but using matrix algebra, computers rapidly perform the task.

The database used here is shown in the appendix to this chapter (table A5.2), where the absolute numbers of animals (rather than percentages) present in each cave is used. Absolute numbers of animals are preferred because percentages may misrepresent the makeup of the animal distribution in caves with a small number of depictions versus caves with large numbers of animals. For example, at the Chabot cave (department of Gard, France) mammoths make up 76 percent of the herbivores, with sixteen mammoths depicted, whereas at Chauvet (Ardèche, France) mammoths make up only 28 percent of the herbivores with a total of sixty-six mammoths depicted, meaning that the Chabot cave would be overrepresented compared to Chauvet using percentages. It is interesting to note that Leroi-Gourhan, in his major work on Franco-Cantabrian cave art[2], when describing animal themes (such as bison-horse, for example), counted themes, so that a single representation counted the same as numerous representations. A more recent study with a larger database follows this same scheme.[3] This method has the potential to lose a lot of information, particularly if the number of representations in any way was intended to reflect the real world. Large numbers of parallel incisions on bones from the Middle Paleolithic suggest some type of accounting system for large quantities already existed at an early stage.

The data in this study are based on thirty-two caves in France and northern Spain, totaling 2,429 animals. Most of the data come from caves that were inventoried in detail by Leroi-Gourhan[2], with the addition of later major discoveries such as Chauvet[4] and Cosquer[5] and a more recent inventory at Rouffinac[6] and Lascaux[7], in addition to the Spanish caves inventoried by J. Altuna.[8] Figure 5.1 shows a frequency histogram of the herbivores in the present study compared to Leroi-Gourhan's summary table, together with the more recent data in Sauvet and Wlodarczyk[3], which includes eighty-four French and Spanish caves and is the largest data set of the three. The three distributions are quite similar, with the horse and bison by far the most common animals depicted, followed by deer, mammoth, and ibex.

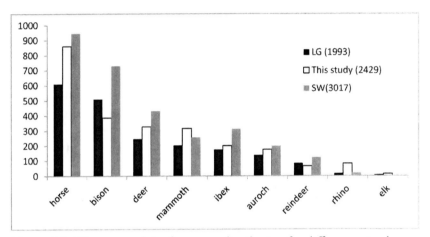

Figure 5.1: Histogram of the frequency distribution for different animals portrayed in French-Spanish cave art using three different data sets: LG–Leroi-Gourhan[2], SW–Sauvet and Wlodarczyk[3] and this study. The size of each data set is indicated. Note the similarity of the three distributions.

Returning to the statistical analysis, the details of which are presented in the appendix of this chapter, the results of the principal component analysis show that 68 percent of the variation observed in the data can be explained by just two variables, with the first dimension (X axis) accounting for 38.4 percent of the variation and the second dimension (Y axis) accounting for 29.4 percent of the variation (figure 5.2, top panel). The animals fall into two distinct groups with no overlap: Group 1 includes the aurochs, horse, red deer, and ibex and shows a strong dependence on the first dimension (X axis) but weak dependence on the second dimension. In contrast, group 2 includes the woolly mammoth, great elk, woolly rhinoceros, reindeer, and bison, and this group shows a strong dependence on the second dimension (Y axis) but a weak dependence on the first dimension.

It is interesting to note that a similar but more complex grouping was obtained by Sauvet and Wlodarczyk[3] who used a similar multivariate statistical technique and looked at a wider range of animals including carnivores (lions and bears) and rarely depicted animals (e.g., birds and fish). In their results, the reindeer, mammoth, and rhino belonged to one group (in common with the results here), and the aurochs and red deer belonged to a second group (also in common with the results here). The ibex, horse, and bison, however, in that study formed a third group between the first and second groups. Curiously, the authors of that study did not propose any physical explanation for their principal components.

We now need to assign a physical interpretation to each of our two dimensions. We might begin by testing the hypothesis of Leroi-Gourhan[2], refuted sometime ago[9,10], that, together with abstract symbols in the caves, the horses, ibex, mammoth, and stags represent a male gender symbolism and that bison, aurochs, and deer (hinds) represent a female symbolism. To test this hypothesis, we could assign a male gender symbolism to the first dimension (X axis) and a female gender symbolism to the second dimension (Y axis), or vice versa. Either way, examination of the two main groups shows that there is no division along these gender lines, even if one accepts the curious assignment of bison to represent a female symbol and the horse a male gender symbol, as proposed by Leroi-Gourhan. The gender of the animals portrayed in cave art is only clear in the case of red deer, where the stag and hind are distinguished by the absence of antlers on the female. Not only does this hypothesis not make any sense, the data do not support it.

Alternatively, we can entertain the hypothesis that one group of animals is more dangerous to hunt but provides greater resources for a given amount of hunting effort (group 2 animals in our case) and that the other group is easier to hunt but provides fewer resources. There is some support for this view insofar as the large mammoth and rhinoceros are part of group 2 and were dangerous to hunt but provided greater resources (for example, hides, bones, and meat) than ibex or red deer because of their mass. This hypothesis, however, boils down to just a single variable—a balance is reached between risk on one hand and reward on the other. A single axis would then explain the majority of the observed variation in a principal component statistical analysis. Clearly, the data require two variables. Although this hypothesis is plausible, the data are inconsistent with this idea.

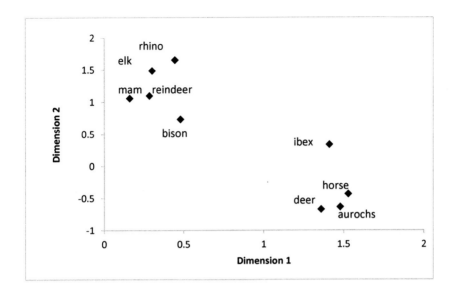

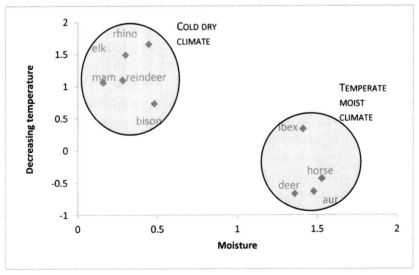

Figure 5.2 Upper panel: Results of principal component analysis. The animals fall into two distinct groups. Group 1 includes aurochs, horse, ibex and red deer. Group 2 includes woolly mammoth and rhino, bison and giant elk. **Lower panel:** The first dimension (X axis) is interpreted as moisture (or aridity), and the second variable (Y axis) is interpreted as decreasing temperature. Group 1 animals therefore represent a temperate-moist climate (interstadial), and group 2 animals represent a cold-dry climate (stadial). These are the two main types of Ice Age climates.

The evidence presented in chapter 3 strongly suggests that the first dimension (X axis) represents aridity or moisture and the second variable (Y axis) represents decreasing temperature (see figure 5.2, lower panel). Group 2 animals, including the woolly mammoth, woolly rhinoceros, giant elk, reindeer, and bison, favor a dry-cold climate typical of stadial periods in a southern tundra-steppe environment (see chapter 4). On the other hand, group 1 animals, comprising aurochs, red deer, and horse, were more suited to a temperate-moist climate in a grassland or open woodland environment, typical of interstadial periods. Important caves with little intermixing of cold-adapted and temperate faunal assemblages that clearly fall into groups 1 and 2 are Lascaux and Chauvet, respectively (see table A5.2 in appendix). These caves would represent end-members in a continuum.

In times of rapid climate change, the two groups would have intermixed to some extent, resulting in mixed assemblages and introducing a substantial random component to the data. This mixing is apparent by perusal of the table in the appendix at the end of the chapter, where the first five columns represent cold-adapted animals, and subsequent columns represent temperate-climate animals. During the Last Glacial Maximum, the archaeological evidence indicates that the woolly mammoth migrated south into Spain, where red deer were common (see chapter 4). The caves at Ebbou, Ardèche, southern France, and Altamira and El Castillo, northern Spain, are dominated by red deer, but all display a single mammoth and appear to be examples of mixed faunal assemblages. There appears to be no good evidence that these lone mammoths were added by artists at a later time, but in the case of El Castillo, it has been suggested that it is in fact a straight-tusked elephant rather than a mammoth. Other examples of this faunal mixed assemblage are Font de Gaume and Pair-non-Pair in France (see appendix, table A5.2).

Leroi-Gourhan[2] noted that the horse-bison and horse-aurochs themes were the most common pairings consistent with the mixed-faunal assemblage proposed here. The bison is clearly a cold-adapted animal (see chapter 4), but the horse is a member of the temperate group in the statistical analysis presented here. At Lascaux, the dominant theme is horse-aurochs, whereas at Chauvet, the pairing is horse-bison. The overall interpretation is that, while horses were cold-adapted, they also tolerated temperate conditions, resulting in mixed faunal assemblages (horse-bison and horse-aurochs) during times of rapid climate change. Table 5.1 summarizes the results of the statistical analysis.

The ibex favors steep rocky slopes and mountainous regions. Ibex are well-represented in caves close to the mountainous regions, such as the Pyrenees, the Massif Central, and the French Alps to the east. The Cougnac cave exhibits thirty six percent ibex and it lies west of the Massif Central. The Cosquer cave, department of Bouches de Rhone, southern France, where 30 percent of the animals are ibex, could reflect ibex from either the Massif Central immediately to the northwest or the Alps immediately to the northeast. Similarly for Montespan, Haute Garonne, situated just north of the Pyrenees, 20 percent of the depictions are ibex. Other examples are Las Monedas and Chimeneas, of the Santander region, northern Spain, west of the Pyrenees, where 13 percent and 33 percent, respectively, of the depictions are ibex. As mentioned previously, ibex-hunting camps have been identified in Solutrean deposits of northern Spain. The Pair-non-Pair cave, however, is a counter example, where 28 percent of depictions are ibex, but it is located near the Atlantic coast, in the department of Gironde. During cold periods, however, the ibex would have been forced onto lower ground by the ice caps. The absence of ibex bones in this region maybe due to lack of preservation during stadial periods when downward erosion was extensive (Chapter 3).

Table 5.1. Summary of Principal Components and Pollen Analyses

Principal components	X axis (moisture)	Y axis (temperature)
Variance	38.4%	29.4%
Climate	Temperate-moist	Cold-dry
Animal assemblage	Aurochs, red deer, horse, ibex	Mammoth, rhino, elk, bison, reindeer
Examples	Lascaux	Chauvet
Vegetation	Woodland	Tundra-steppe
Pollen	Trees dominant	Grasses/sage dominant

The idea that the Chauvet cave faunal assemblage is typical of a stadial (cold) period and that the Lascaux cave art represents an interstadial (temperate) period can be tested by looking at pollen assemblages in these two end-member caves. The study of pollen, called palynology, is an important part of paleoclimatology studies.[11] Palynological studies of a marine core taken off the Atlantic coast of southern Spain and a core taken from the Mediterranean Sea show very good correlation with climate changes indicated by the European (GRIP) and American (GISP2) ice cores from Greenland.[12,13]

Plants spread their pollen by various mechanisms in order to reproduce—wind being the most important mechanism of dispersion, but water and insects, bees for example, also spread pollen. Pollen grains are very small, about 10–100 millionths of a meter, and are usually counted and identified under a microscope. They are also very resistant to harsh chemical environments and so are commonly well-preserved over a long time period. Because pollen from different plant species has distinct morphological features, such as size and shape, the ancient pollen record provides a record of the vegetation in past times. This record is often preserved in lake sediments, ocean sediment cores, and peat bogs. Peat bogs became common in the Holocene, so the pollen record for that period is particularly good. If the sedimentation rate is relatively fast, in lake sediment for example, the pollen record will provide a snapshot of the vegetation for a relative short time frame in the past, and a snapshot of a longer time frame if the sedimentation rate is slower, such as in ocean sediment cores. To assign the assemblage of pollen grains observed in a sample to a given time period, the age of the sediment obviously must be known, and this age constraint is often provided by carbon-14 dating of organic material in the sediment. Whether transported by water or wind, the pollen in a sample may or may not be representative of the original vegetation, due to differential transport of smaller versus larger grains. In addition, different species of plant produce different amount of pollen, and this must be taken into account in any reconstruction of past vegetation. Any disturbance of the pollen-bearing sediment by water currents or burrowing animals can distort the pollen signal.

Pollen in cave sediments that contain cave art may provide valuable insight into the climate at the time the artwork was undertaken. Caves, however, are commonly subjected to flooding, which may disturb or obliterate the stratigraphic record or introduce new pollen from a different time frame, giving rise to a false picture of the vegetation at that time. In addition, animals (such as cave bears) or humans may bring pollen on their fur or clothes into caves. Moreover, modern pollen transported by air currents or water could swamp any ancient pollen signal.

This makes interpretation of cave pollen a difficult task. A relatively simple pattern, however, is commonly observed in pollen studies of the last glacial period—pollen associated with herbaceous trees (for example hazel, elm, birch, beech) are relatively common during temperate periods, indicating a wooded landscape, whereas tree pollen is relatively rare in cold climates (and if present at all, it is commonly pine), where grasses and sage dominate.[11,12]

The Chauvet cave was discovered in 1994 in the department of Ardèche, in southwest France, and named after one of its three discovers, who were speleologists. Under the direction of prehistorian Jean Clottes, French officials were determined not to allow a repeat of previous mistakes made after other important cave art discoveries (Lascaux, for example). The public was prohibited from entering the cave, and the humidity and temperature of the cave were controlled. Moreover, few professional cave art experts were allowed to visit the cave—and then only for short periods of time and under strict guidelines. Narrow metal walkways were established throughout the cave, from which all studies were to be undertaken remotely, without touching the walls or walking on the floor.[14] These stringent conditions make the Chauvet cave an ideal place for palynological studies. A documentary film inside the cave was released in 2010 by director Werner Herzog, entitled *The Cave of Forgotten Dreams* — it used 3-D photography.

Pollen was studied in floor sediments in several areas of the cave that were above the flood line and far from the cave entrance, to avoid modern pollen contamination. The results are consistent among isolated samples and samples taken in stratigraphic order.[15] The dominant pollen present were grasses (*Graminae*) and sage or wormwood (*Artemisia*), typical of a dry steppe environment. Tree pollen made up only 1.5 to 11 percent of the pollen, consisting of cold-adapted trees like pine, indicating an open grassland environment. Only 1 or 2 percent of the pollen was of modern origin indicating little contamination. Remnants of burned pine are present in hearths throughout the cave, indicating a source of pine wood was available locally, but it is poorly represented in the pollen spectrum. These preliminary pollen data support a dry-cold period corresponding to the present occupation level, consistent with the idea that Chauvet was painted during a stadial period and also consistent with the cold-adapted faunal assemblage depicted there.

In the case of the Lascaux cave, discovered in 1940, it was visited by the public until it was closed in 1963. Early pollen studies did not produce sufficient pollen for reliable results. Later sampling in 1975 using newer techniques, however, produced good results.[16] The layer that was analyzed for pollen also contained ocher powder used to make some of the paintings and also yielded chert tools, providing a connection between the Magdalenian painters and the pollen.[17] Thirty percent of the pollen consisted of maritime pine trees (*Pinus pinaster*), as well as broadleaf trees, such as hazelnut, walnut, and oak, and 20 percent grasses (*Graminae*), indicating an open temperate forest referred to as the Lascaux interstadial.[16] This layer is dated at about 17 ka ^{14}C BP (19 ka cal BP) from burned charcoal fragments and occurs as a short warm period during the Last Glacial Maximum. The Lascaux interstadial is also recognized at other locations of the same age in France and Spain; it is also present in the oxygen isotope record in Greenland ice cores[17], but it is not recognized on the methane record shown in figure 3.1. The length of the Lascaux warm period may have been too short for methane sources, such as wetlands and bogs, to respond to the rapid change. The pollen record is consistent with the idea that the Lascaux cave was painted during this brief warm period.

This still leaves unexplained the apparent enigma, already outlined in chapter 1, of the dominance of reindeer bones found in the archaeological levels at Lascaux, though only one reindeer is depicted in the wall art. A radiocarbon date on a reindeer antler from the shaft at Lascaux (18,600 ± 190 ^{14}C BP)[18] yields a calibrated age range of 22.0–22.9 ka cal BP—too old for the Lascaux interstadial. The artwork at Lascaux remains poorly dated, and because of the absence of carbon in the paint used, together with the fact that the archaeological excavations were done under less-than-ideal conditions, will probably remain so for the foreseeable future.[19] In other words, the reindeer bones and the cave art are probably not of the same age (see chapter 7).

In summary, the main conclusions of this chapter are that the herbivores in cave art fall into two distinct groups using multivariate statistics (principal component analysis) and that over two thirds of the variance observed in the animals depicted can be explained by just two Ice Age climate variables, namely temperature and aridity. One group of herbivores is adapted to a cold-dry climate (woolly mammoth, woolly rhinoceros, reindeer, bison and giant deer) and one group is adapted to a moist-temperate climate (red deer, aurochs, horse and ibex). In the next chapter, we will examine stylistic and radiometric methods for dating cave art and also the techniques used in producing the cave imagery.

Appendix: Principal Component Analysis

The data in table A5.2, representing thirty-two caves and nine herbivores (corresponding to thirty-two cases and nine variables), were analyzed using IBM's statistical package SPSS 22. Two statistical tests are commonly applied before subjecting data to principal component analysis to determine whether the data have an underlying structure and are not simply random. The first test is Barlett's test of sphericity, which is also related to the chi-squared (χ^2) test of significance. The second test is the Kaiser-Meyer-Olkin (KMO) measure of sampling adequacy.[20] In the case of Bartlett's test, small values (< 0.05) and large chi-squared values (greater than those predicted for a random sample) indicate that factor analysis can be usefully employed. The data set used here yield a value of 0.0 for Bartlett's test and a chi-squared value of 201, indicating that the data are not random. For the KMO test, the data yield a value of 0.63, which, on a scale of 0 to 1, is above the suggested minimum value of 0.5, indicating again that the data are not random.[20] A third test is to compare the results from our data with randomized data sets. Table A5.1 shows the results from five randomized data sets (using a random number generator) with the same size as that in our dataset. The results of Bartlett's test, the KMO test, and the chi-squared test are also shown. The randomized data have distinctly different statistics compared to table A5.2, including lower KMO values (<0.51), higher Barlett's test values, and lower Chi-squared values than our data set. The results in table A5.1 indicate that principal component analysis is justified on the bases of these three statistical tests.

Table A5.1. Statistical Tests

	Bartlett's	KMO	$\chi 2$
Table A5.2	0.0	0.63	201
Random 1	0.97	0.51	16
Random 2	0.71	0.51	23
Random 3	0.25	0.39	33
Random 4	0.13	0.41	36
Random 5	0.27	0.37	32

The results of the principal component analysis shows that 68 percent of the variation observed can be explained by two variables, with the first principal component accounting for 38.4 percent and the second component 29.4 percent of the variation. By comparison, the first two principal components could only explain 39 to 46 percent in the case of the random data sets in table A5.1. This is additional evidence that these results are statistically valid. So that these results can be replicated, the following options were used in SPSS 22 under the categorical principal component analysis package: variable weight (1); variable scaling (numeric); variable grouping (multiplying); options (exclude objects with missing values); variable principal (object principal).

Table A5.2. Frequency of Herbivores in French/Spanish Caves

Raw numbers	ReinD	Mam	Rhino	Elk	Bis	Ibex	Hor	Aur	RedD	Tot
Niaux	0	0	0	0	36	10	18	2	2	68
Altamira	8	1	0	0	27	11	30	5	35	117
Cougnac	0	6	0	3	0	8	1	0	4	22
Pech Merle	0	7	0	1	7	1	6	3	2	27
Rouffignac	0	158	11	0	28	12	16	0	0	225
Le Portal	0	0	0	0	20	2	28	0	4	54
Chauvet	14	68	66	5	32	20	51	9	2	267
Cosquer	0	0	0	2	10	28	65	7	17	129
Las Monedas	4	1	0	0	2	4	14	0	1	26
Chimeneas	0	0	0	0	0	5	6	8	11	30
Lascaux	1	0	2	0	20	35	355	87	88	588
Combarelles	8	7	1	0	16	2	20	2	0	56
Font de Gaume	0	23	1	0	51	0	26	6	14	121
Pair-non-Pair	0	2	0	1	2	7	6	3	4	25
Gargas	0	4	1	1	30	4	23	3	5	71
La Pasiega	0	0	0	0	7	7	39	14	50	117
Le Gabillou	5	0	0	0	9	6	28	9	0	57
Villars	1	0	0	0	3	2	11	1	0	18
Pergouset	0	0	0	1	3	3	3	0	3	13
Cheval	2	8	0	0	3	0	2	0	0	15
Covalanas	0	0	0	0	0	0	1	0	19	20
Teyjat	5	0	0	0	3	0	7	3	15	33
Trois Freres	4	2	0	0	6	1	7	0	0	20
Montespan	0	0	0	0	0	5	8	3	8	24
Ebbou	0	1	0	0	2	9	16	5	15	48

Raw numbers	ReinD	Mam	Rhino	Elk	Bis	Ibex	Hor	Aur	RedD	Tot
Oulen	0	5	0	0	2	1	0	0	0	8
Chabot	0	16	0	0	0	0	4	1	0	21
Baume Latrone	0	7	0	0	0	0	1	0	0	8
Le Mas D'azil	0	0	0	0	9	0	5	0	3	17
Tito Bustillo	7	0	0	0	3	9	27	3	23	72
Altxerri	6	0	0	0	44	5	2	2	0	59
Ekain	0	0	0	0	11	5	34	0	3	53
Totals	**65**	**316**	**82**	**14**	**386**	**202**	**860**	**176**	**328**	**2429**

References cited

1. Rencher, A. C. and Christensen, W. F., 2012. *Methods of Multivariate Analysis*. Third ed., Wiley, New Jersey.
2. Leroi-Gourhan, A., 1967. *Treasures of Prehistoric Art.*, Abrams Press, New York.
3. Sauvet, G. and Wlodarczyk, A., 2008. "Towards a formal grammar of the European Paleolithic cave art." *Rock Art Research*, 25, 165–172.
4. Clottes, J., Gély, B. and Le Guillou, Y., 1999. "Complementary iconographic information from the Chauvet Cave." *International Newsletter on Rock Art*, 24, 5–8.
5. Clottes, J. et al., 2005. *Cosquer redécouvert*. Seuil, Paris.
6. Plassard, J., 1999. *Rouffignac. Le Sanctuaire des Mammouths*. Seuil, Paris.
7. Aujoulat, N., 2005. *Lascaux: Movement, Space, and Time*. Abrams Press, New York.
8. Altuna, J., 1983. "On the relationship between archaeo-faunas and parietal art in the caves of the Cantabrian region." *Animals and Archaeology*, vol. 1, *Hunters and their prey*, 227–238. Clutton-Brock, J, and Grigson, C., eds. British Archaeological reports, International Series, 163.
9. Ucko, P. J. and Rosenfeld, A., 1967. *Paleolithic Cave Art*. World University Library.
10. Bahn, P. G. and Vartut, J., 1997. *Journey Through the Ice Age*. University California Press, Berkeley.
11. Bradley, R. S., 1999. *Paleoclimatology, reconstructing the climates of the Quaternary*. Second ed., Elsevier Academic Press, New York.
12. Nebout, N. C. et al., 2002. "Enhanced aridity and atmospheric high-pressure stability over the western Mediterranean during North Atlantic cold events of the past 50 ka." *Geology*, 30, 863–866.

13. Sánchez Goi, M. F. et al., 1999. "High resolution palynological record off the Iberian margin: Direct land-sea correlation for the last interglacial complex." *Earth and Planetary Science Letters*, 171, 123–137.
14. Baffier, D., 2005. "La Grotte Chauvet: conservation d'un Patrimoine." *Bull. Soc. Préhistorique Francaise*, 102, 11–16.
15. Girard, M., 2005. "Analyses polliniques des sols Aurignacians de la Grotte Chauvet (Ardèche). Resultats Préliminaires." *Bull. Soc. Préhistorique Francaise*, 102, 63–68.
16. Leroi-Gourhan, Arl. and Girard, M., 1979. *Analyses polliniques de la grotte Lascaux in Lascaux Inconnu*, (Leroi-Gourhan, Arl. and Allain, J. eds). CNRS, Paris.
17. Leroi-Gourhan, Arl., 1997. "Chauds et froids de 60,000 a 15,000 BP." *Bull. Soc. Préhistorique Francaise*, 94, 151–160.
18. Valladas, H. et al., 2013. "Dating French and Spanish prehistoric decorated caves in their archaeological contexts." *Radiocarbon*, 55, 1422–1431.
19. Delluc, B. and Delluc, G., 2006. *Discovering Lascaux*. Éditions SudOuest.
20. Dziuban, C. D. and Shirkey, E. C., 1974. "When is a correlation matrix appropriate for factor analysis?" *Psychological Bulletin*, 81, 358–361.

6 Techniques and Dating of Cave Art

The techniques used by the Upper Paleolithic cave artists were relatively simple and few in number. It is useful to review them before discussing how cave imagery is dated. The most important techniques in cave art are painting, drawing and engraving, with sculpture being the least common technique. One of the best-known sculptures in soft clay is the pair of bison leaning on a limestone block on the cave floor at Le Tuc d'Audoubert (Ariège), but such skillful examples are rare.[1] Painting and engraving (with a sharp stone tool), are by far the most common techniques used. At Rouffignac (Dordogne), for example, about two-thirds of the figures are engraved, and the rest were drawn in black. Other, less commonly used, colors are red and yellow. The raw source minerals for these colors were hydrated iron oxides (mainly limonite or goethite) for yellow, iron oxide (hematite) for red, and manganese dioxide (pyrolusite) for black. Charcoal was also used for black drawings allowing the images to be radiocarbon dated. Drawing was done with a piece of charcoal or other mineral substance used as a crayon on the wall surface, and several such mineral fragments showing abrasion marks were found in the shaft at Lascaux.[2] Overall, the raw materials (iron and manganese oxides and charcoal) were widely available near the caves, and these resources appear to have been plentiful.

Detailed analyses, using electron microscopy and X-ray analysis, show that the basic paint colors were made from ground hematite and manganese oxide mixed with water, but they also contained additional ingredients. At Cougnac (Lot, France), work by M. Lorblanchet and his colleagues has shown that additives to the red ocher included quartz, clay minerals, and biotite.[3] These additives acted as binders and extenders of the paint. Similar "recipes" were identified at Niaux (Ariège, France), where feldspar (recipe F), biotite (recipe B), and talc (recipe T), were also added to manganese oxide and charcoal for the black figures in the Salon Noir.[4] The raw materials were all available locally. Directly opposite Niaux is the La Vache cave, where detailed analysis on portable art works show that recipe B was used for ocher-stained portable artifacts.[5] The proximity of Niaux and the similar paint recipe at La Vache[5] across the valley suggests the painters of at least part of Niaux inhabited the cave at La Vache.[4]

At Chauvet (Ardèche, France), a combination of techniques was used for the famous panel of the horses.[6] The cave wall consisted of white calcite covered with a thin layer of soft clay. The wall was prepared by rubbing off the clay layer to reveal a white patch. A preliminary outline was done with a sharp tool. Charcoal outlines of the horse were drawn and filled in with charcoal shading by rubbing. Final touches for the head and mane were made with a sharp tool. These simple techniques were combined in various sequences for some of the other panels. Work by N. Aujoulat at Lascaux also indicates that some figures were also outlined by engraving, with the engraving typically being less than a centimeter in width and depth, and later overprinted by painting using manganese oxide.[7] This technique appears to have been used on more friable parts of the wall, whereas painting was used on more indurated surfaces. Painting can be done by brush strokes that provide relatively narrow lines for an outline, or by daubing paint with a piece of fur or leather to fill in a figure with shading. Closely spaced dots of black paint were also used to outline parts of some figures at Lascaux, presumably by dipping a brush in paint pigment. Spray painting, either directly from the mouth or through a straw or hollow bone, also appears to have been used, producing a more diffuse pattern, for an animal's mane or tail, for example. This technique could only be used in cases where the painter could get within inches of the wall. Some type of scaffolding could accomplish this, but more often, painting by brush, possibly on an extended stick, was more likely. In cases where the outline of part of an animal's body is quite sharp, a mask or stencil of some kind was used. At Lascaux, the horses were painted first followed by aurochs and stags were painted last. This sequence is observed in several panels suggesting a pre-planned composition for these works.[7]

People's hands were extensively used as stencils in many caves, where the hand was placed on the wall palm down, and paint was sprayed on the hand, producing negative prints. Positive hand prints could be produced by putting paint directly on the hand and then placing it on the wall. The size and dimensions of some of these hands suggest they were commonly of women and children.[8] As noted above, drawing done with charcoal can be dated directly using the carbon-14 technique. Dating of cave art by stylistic criteria and by the carbon-14 method is discussed below.[9]

The stylistic method is a technique in which cave art is assigned to a cultural period (for example, early or late Magdalenian or Solutrean) based on the style of execution. Various "signs" or abstract symbols that commonly accompany the animals have also been used in stylistic determinations. The meaning of these symbols, however, remains a mystery.

Radiometric methods, on the other hand, assign a specific time frame in years before the present to the paintings based on radioactive decay of an element (e.g., carbon or uranium). Radiometric dating can be of two types: direct dating of the painting itself using samples from the cave wall or indirect dating by association with dated objects, such as charcoal from a nearby hearth or a bone tool. In some cases, rock falls of wall paintings become buried in archaeologically dated horizons in a cave, placing a minimum age on the artwork, which cannot be younger than the sediments in which it is embedded. Also, similarities between buried decorated objects with the cave wall artwork can be used to date a painting. In rare cases, a major rock fall has sealed off a cave (such as Chauvet), placing a minimum age on the paintings if no other entry was possible.[10]

The best-known radiometric dating method is the carbon-14 method, developed by W. F. Libby and his colleagues at the University of Chicago in the 1950s. It continues to be the dating method of choice in the archaeological world. This method was first successfully tested on objects whose age was already known, such as a 3,000-year-old redwood tree and 5,000-year-old wood from an Egyptian tomb. Libby received the Nobel Prize in chemistry in 1960 for this work.[11] Whether the technique would also work on older samples was unknown at the time. The first radiocarbon date on an unknown sample was on charcoal from Lascaux (from the shaft), and it was dated by the Chicago group in 1951. It yielded an age of 15.5 ka ^{14}C, with a rather large error, by today's standards, of ± 900 years. This age was un-calibrated, since it was not yet known that radiocarbon ages, especially ages this old, needed calibration (see below). With modern calibration, the Lascaux date lies in the range of 16.5 ka to 21.0 ka cal BP, which places the Lascaux charcoal in the late Solutrean to early Magdalenian cultural periods which is a large time interval.

Abbé Breuil (1877–1961) was the first paleohistorian to study Lascaux after its discovery in 1940, and during his long career, he studied most of the known cave art in France and Spain by tracing the images directly from the walls—he himself was a talented artist. He had come to believe, as he described in his book *Four Hundred Centuries of Cave Art*, that Lascaux was Gravettian in age (Perigordian in the older terminology), based on the similarity of Lascaux drawing style to locations where the artwork was known to be late Aurignacian or Gravettian. The Lascaux bulls (or aurochs) are distinctive in that their horns were painted in what Abbé Breuil called twisted perspective. The animals themselves are usually drawn in profile, but the horns have a more frontal view, sometimes called half or three-quarters perspective. The older localities had the same style as the Lascaux bulls, so Abbé Breuil respectfully rejected the radiocarbon date as being too young.[12]

This disagreement between dating of cave art based on style versus radiometric dating was the first such example, but not the last. Today, a similar debate continues between stylistic dating and radiometric dating in the case of the Chauvet cave (see chapter 2). Instead of arguing against a single radiometric date on a sample from Lascaux, as Abbé Breuil did, the current stylists must contend with over 100 radiometric dates on charcoals from ten different international laboratories yielding concordant results, and also with direct dates from the drawings themselves, which is a much higher bar to cross. Abbé Breuil turned out to be incorrect, and modern dates on Lascaux bones and charcoal are similar, within error, to Chicago's first date (see Chapter 7). The debate over the age of Chauvet between the stylists and those who favor the older radiometric dates continues today.

The groundwork for stylistic dating was laid by the two scholars, Abbé Breuil and André Leroi-Gourhan (1911–1986). Leroi-Gourhan occasionally agreed and also respectfully disagreed, with some of Breuil's earlier stylistic age assignments. Breuil proposed two independent cycles of cave art: an earlier Aurignacian-Gravettian cycle and a later Solutrean-Magdalenian cycle.[13] Leroi-Gourhan proposed four styles (stages I through IV) with a progressive evolution over time, with largely gradational and poorly defined changes between stages, making it a somewhat subjective scheme.[14] Style

I covered the Aurignacian cultural period, style II covered the Gravettian and Solutrean periods, and styles III and IV covered the Magdalenian, the latter two styles accounting for 80 percent of the caves he examined. Leroi-Gourhan's style scheme has largely replaced Breuil's scheme. Because there is no sharp boundary between the styles, it can be difficult to assign a particular painting to a particular style, and Leroi-Gourhan himself provided examples of different styles, such as the way the horns of an ibex were portrayed in the same cave[14], recognizing that style cannot always reliably date a painting or engraving.

For style criteria to be diagnostic, one would have to know which techniques and conventions persisted where and for how long—and that these conventions did not recur through time.[15] As cave art is still not understood, we do not yet have that information. In addition, each time a new cave is discovered, old ideas often require revision. Leroi-Gourhan pointed out that horns in twisted perspective, as on the bulls at Lascaux, are not confined to Breuil's first cycle (Aurignacian-Gravettian), permitting Lascaux to be Magdalenian or Solutrean in age, as indicated by the first charcoal radiometric age. Indeed, the discovery of the Chauvet cave in 1994 and its radiometric dating to the Aurignacian cultural period made most specialists reconsider the idea that cave art followed a progressive increase in technical expertise and sophistication with time. Almost fifty years ago, the point was made that it is unreasonable to expect a large body of work lasting 25,000 years to follow a simple progression from simple to complex, but rather there would be false starts, dead ends, and climaxes throughout this long period.[16]

Those scholars unwilling to reconsider traditional views for the development of cave art (i.e., a linear increase in sophistication and technique with time) favor a stylistic dating approach over radiometric dating, as in the case of Chauvet. Using stylistic criteria one author[17] believes that the black animal depictions at Chauvet are Solutrean to Magdalenian in age, rather than Aurignacian, based on similarities of several animals at Chauvet to those at Lascaux, for example. The similarities are indeed present, but these similarities cannot by themselves negate the large body of internally consist radiometric data for the Chauvet cave.

A more recent study of the stylistic dating technique shows that it can be based on circular reasoning, whereby sites are dated stylistically by comparison to other caves also dated by style—without reference to radiometrically dated sites.[18] Since only a small percentage of sites will ever be dated radiometrically (because of the paucity of suitable carbon-based samples on the paintings), a way forward suggested by these workers and others is to use radiometrically dated sites as "anchor points," together with stylistic considerations, to better date cave art.[19]

Rumors of the death of stylistic dating, in the case of the Côa Valley petroglyphs, Portugal, appear to be premature.[20] Three different dating techniques indicated a Holocene age rather than a Paleolithic age for one cave in Côa Valley. Subsequent work showed, however, that the carbon-14 dates were in error because of the extremely small amounts of carbon present on the rock engravings.[21] Many of the petroglyphs at Côa Valley are now thought to be Upper Paleolithic in age, based mainly on stylistic criteria. Several examples do exist where radiometric dates and stylistic considerations agree with one another, and one of these caves (Altamira) is discussed later in the chapter. First, the carbon-14 dating method is described in more detail below.

Most carbon (99 percent) has an atomic weight of twelve units (denoted as ^{12}C). A small amount of carbon is slightly heavier, with a weight of fourteen, ^{14}C; this carbon is radioactive. It is produced by interaction of cosmic rays with nitrogen in the atmosphere, and the ^{14}C produced then begins to slowly decay back to nitrogen over time. Trees and animals absorb this carbon-14 from the atmosphere and hydrosphere. The carbon-14 reaches a constant level in living organisms because it is simultaneously being absorbed and decaying. However, when the plant or animal dies, it stops absorbing the carbon-14, and the carbon-14 begins to decay without being replenished. It takes about 5,700 years for the carbon to decay to half its original size (the so-called half-life of carbon-14). By measuring the amount of carbon-14 in a piece of wood or animal bone, we can estimate the length of time since the death of the animal or plant. This is the basis of the carbon-14 dating technique.

How far back in time is the carbon-14 technique useful? If you try to fold a sheet of paper in half several times, it becomes very difficult after about six folds, as the paper gets progressively smaller. Analogously, the carbon-14 dating technique extends back in time only about eight times the half-life (8 x 5,700), or about 45,000 years, which extends back to the Aurignacian cultural period. The amount of carbon-14 present today in such an old sample is so small that it is difficult to measure against background radiation. In addition, small amounts of contamination from younger or modern carbon can give an age that is much too young. Using modern techniques and equipment, the carbon-14 technique is not reliable beyond about 50,000 years into prehistory. This is within the expected age range of most European cave art.

One of the assumptions of the carbon-14 technique is that the amount of carbon-14 produced in the atmosphere is constant over time. It turns out that this assumption is incorrect for reasons we need not go into here. But it means that carbon-14 dates must be corrected, usually by increasing the age by about 10–20 percent (see, for example, table 2.1). The amount of calibration must be determined by independent dating techniques, such as counting tree rings or using the uranium-thorium technique for dating corals and calcite cave deposits.[22] For example, a carbon-14 age of 15,000 years BP would be about 18,200 years BP when calibrated using other independent techniques. As pointed out in chapter 2, the raw carbon-14 age is written as 15.0 ka ^{14}C BP and as 18.2 ka cal BP when calibrated, where BP stands for "before present."

It is further assumed that radiocarbon results have a bell shaped curve or normal distribution. The error associated with each radiocarbon date can be quoted at the 68 percent confidence level (one standard deviation or 1σ) or at the 95 percent level (two standard deviations or 2σ). In other words, the probability that the actual age of a sample lies within the stated range is either 68 percent or 95 percent. For example, the radiocarbon age 15.0 ka ± 200 ^{14}C BP, would calibrate to 18.0 ka–18.5 ka cal BP (1σ) or 17.8 ka–18.7 ka cal BP (2σ).[23] The higher confidence interval (2σ) has a wider age range. Most of the dates quoted in this book are at the 2σ confidence level.

Calibrated carbon-14 dates can be compared directly with the ice core climate curve discussed in chapter 3. Since climate change occurs on a millennial or shorter time scale (see chapter 3), un-calibrated carbon-14 dates will usually not correspond to the climate at that time, given typical error bars and calibration corrections. For this reason, much of the older archaeological literature related to climate change using un-calibrated dates is suspect, and this is probably one of the main reasons why paleohistorians of cave art have not had much success in relating cave art to climate change. An exception is the paper by a French group[24]; these workers showed, using calibrated carbon dates and the Greenland ice core chronology, that there was a hiatus in cave imagery production during the Last Glacial Maximum (Heinrich event 2), possibly because the caves were frozen.

In conventional carbon-14 dating, the amount of carbon-14 is measured by counting the radioactive emissions (beta particles), and this requires a large amount of sample (several grams). In the past few decades, a new technique called accelerated mass spectrometry (AMS) has been developed whereby the different carbon isotopes (^{12}C, ^{13}C, ^{14}C) in the sample are measured by accelerating these ions into a strong magnetic field along a curved path that allows separation of the different isotopes according to their weight. The heavier isotope is pushed to the outside curve where the number of heavier ^{14}C ions is counted, so that only a very small sample size is required (thousandths of a gram). This technique allows cave art drawn with charcoal or carbon-bearing pigments to be directly dated without damaging the artwork. This is the direct radiometric dating method referred to earlier.

The small sample size used in AMS-^{14}C dating is an advantage when it comes to dating cave art, but it also has the disadvantage that equally small amounts of contamination by older or younger carbon can result in ages that are too young or too old, respectively. For example, a sample 30,000 years old, if contaminated with only 2 percent modern carbon, would yield an age of 25,000 years, which is a large error.[25] If the sample originally consisted of one one-thousandth of a gram (1 mg) of carbon, 2 percent contamination would only amount to 0.02 mg, equivalent to a few specks of carbon. This is why modern dating laboratories must be cleaner than any hospital operating theatre. A recent study estimated that 75 percent of all carbon-14 dates published only a decade or so ago are too young, due to contamination from modern carbon.[26] This contamination can take place in the laboratory, during sampling in the field—or at any time after the animal or plant died.

In the laboratory, after chemical treatment of the sample by acidic and alkaline solutions, two fractions are commonly obtained—a charcoal (pure carbon) fraction and a humic (organic acid) fraction. If both fractions yield similar ages, then contamination may not have been a major problem. Sometimes, however, the two fractions yield substantially different ages within the statistical error, making the possibility of contamination with younger carbon likely, making the age too young.

In the case of Chauvet, however, the argument is that the ages are too old, not too young, so it has been suggested the samples are contaminated by *older* carbon.[27] The cave contains a large number of bear bones; the bears frequented the cave as far back as 37,000 years ago. If the charcoal from the hearths that make up most of the Chauvet charcoal dates consisted of charcoal from trees mixed with a large fraction of burned bear bones, then a substantial amount of older carbon may contaminate the samples, making the ages appear older.[27]

One way to distinguish between charcoal from plants and that from burned bones is to examine the $^{13}C/^{12}C$ ratio (the so-called delta ^{13}C value) in the samples. One of the international laboratory reports on Chauvet samples reported these delta values.[28] Their delta ^{13}C values were consistently about minus 23 from several different laboratories, which is similar to coal, which typically has a delta ^{13}C value of minus 25. Wood charcoal should be similar to coal because both come from plant matter. On the other hand, a recent study of Aurignacian bones show delta ^{13}C values of minus 20 or greater (-18 to -20)[29], a very significant difference, meaning the bones are richer in ^{13}C than wood charcoal. These data rule out the

Figure 6.1: Part of the Altamira ceiling. Radiometric dates for two numbered polychrome bison are shown in figure 6.2. The total length of the entire painted ceiling is 14 meters (46 feet). After Sautuola (1880).[31]

possibility that the Chauvet samples analyzed were mostly of bone origin, as required by the bear bone contamination hypothesis described above.[27] An earlier study of the charcoal at Chauvet indicates it is entirely wood, mainly pine.[30]

It is interesting to note the cases where the artwork has been dated both radiometrically and by stylistic criteria and in which the methods agree with one another. The probability of this happening by chance is 50 percent, the same as flipping a coin. Important cave art in which this is the case include the Magdalenian caves of le Portel, Niaux, and Altamira. The Altamira cave, in the Cantabrian province, Spain, presents an interesting case of radiometric and stylistic dating, and the evidence is outlined below in more detail.

Altamira is located in northern Spain, about 5 km from the coast. Discovered by a hunter in 1868, Altamira was not examined until 1879, by the Spanish archaeologist Sautuola, and it was not recognized as Paleolithic in age until 1902, when the eminent prehistorian of cave art Emile Cartailhac changed his mind and accepted its antiquity. The cave artwork was first studied in detail by Cartailhac and Breuil in 1905. Archaeological excavations indicate Solutrean and Magdalenian occupation time periods. Both Solutrean and Magdalenian deposits contain bones of red deer, seashells, and fish. Also excavated were Magdalenian bone shoulder blades engraved with the heads of female deer, characterized by distinctive fine scratches on the bone.[32] Stylistically identical deer are present on the walls of the cave itself, leaving no doubt in Henri Breuil's mind that they were contemporaneous. A piece of red ocher also ties the archaeological deposits to the age of the cave paintings. One of the most spectacular of all Paleolithic cave art depictions is the famous ceiling composition of Altamira, consisting of twenty-four lifelike images, located near the cave entrance and composed mainly of bison in various poses—crouching on the ground, leaping, and standing still (see figure 6.1). Many of these images are polychrome bison undertaken in yellow, brown, and red colors.

There are several direct radiometric carbon-14 dates on bison on the Altamira ceiling, in addition to that of the engraved shoulder blade bones decorated with a female deer from a Magdalenian horizon, mentioned above. These data are shown in figure 6.2. They include two sets of data from the same laboratory published in 1992[33] and 1997[34]. The carbon fractions are shown on the left-hand side, and the humic fractions on the right-hand side of the diagram. In the case of the black bison number forty-four (not shown), the humic fraction is somewhat older, suggesting some contamination. An alternative interpretation of the carbon fraction data is that bison forty-four (painted in black), is actually younger than bison number thirty-three and bison number thirty-six, which are polychrome figures (see figure 6.1). Radiometric dates of cave imagery cannot therefore be taken at face value. The calibrated age range for the data shown is 18.4–15.4 ka cal BP, corresponding to the middle Magdalenian. The similar ages of the female deer engraved on a bone fragment and the stylistically similar female deer on the wall art are consistent with a Magdalenian age. Altamira is an example of where both stylistic evidence and the radiometric ages agree with one another. Dating cave art is therefore not an either or proposition (style versus radiometric dates) but both approaches are required.

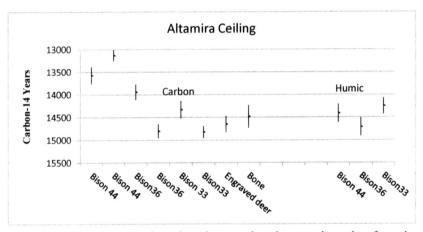

Figure 6.2: Plot of direct radiocarbon dates on three bison and one deer from the ceiling and a date on an engraved bone at Altamira, Spain. Analyses of both carbon (left) and humic fractions (right) are shown. For bison 44 differences between carbon and humic fraction ages could be due to contamination.

References cited

1. Clottes, J., 2008. *Cave Art*. Phaidon Press.
2. Ruspoli, M., 1989. *The cave of Lascaux*. Abrams, New York.
3. Lorblanchet, M. et al., 1990. "Paleolithic pigments in the Quercy, France." *Rock Art Research*, 7, 4–20.

4. Clottes, J. et al., 1990. "La préparation de peintures Magdaléniennes des cavernes ariégeoises." *Bull. Soc. Préhistorique Francaise*, 87, 170–192.
5. Buisson, D. et al., 1989. "Les objects colorés du Paléolithique supérieur: cas de la grotte de La Vache (Ariège)." *Bull. Soc. Préhistorique Francaise*, 86, 183–191.
6. Tosello, G. and Fritz, C., 2005. "Les dessins noirs de la grotte Chauvet-Pont-d'Arc: essai sur leur originalité dans le site et leur place dans l'art Aurignacien." *Bull. Soc. Préhistorique Fracaise*, 102, 159–171.
7. Aujoulat, N., 2005. *Lascaux: Movement, space, and time*. Abrams, New York.
8. Guthrie, R. D. 2005. *The Nature of Paleolithic Art*. University of Chicago Press.
9. Pettit, P. and Pike. A., 2007. "Dating European Paleolithic cave art: Progress, prospects, problems." *J. Archaeological Method and Theory*, 14, 27–47.
10. Sadier, B. et al., 2012. "Further constraints on the Chauvet cave artwork elaboration." *Proceedings of the National Academy Sciences*, USA, 109, 8002–8006.
11. Libby, W. F., 1980. "Archaeology and radiocarbon dating." *Radiocarbon*, 22, 1017–1020.
12. Breuil, H., 1954. "Les Datations par ^{14}C de Lascaux (Dordogne) et Philip cave (S.W. Africa)." *Bull. Soc. Préhistorique Francaise*, 51, 544–549.
13. Breuil, H. 1952. *Four Hundred Centuries of Cave Art*. Translated by Mary E. Boyle. Hacker Art Books, New York.
14. Leroi-Gourhan, A., 1967. *Treasures of Prehistoric Art*. Abrams, New York.
15. Clottes, J., 1996. "The Chauvet cave dates implausible?" *International Newsletter of Rock Art*, 13, 27–29.
16. Ucko, P. J. and Rodenfeld, A., 1967. *Paleolithic Cave Art*. World University Library.
17. Zuechner, C., 1996. "The Chauvet cave: Radiocarbon versus archaeology." *International Newsletter of Rock Art*, 13, 25–26.
18. von Petzinger, G. and Nowell, A., 2011. "A question of style: Reconsidering the stylistic approach to dating Paleolithic parietal art in France." *Antiquity*, 85, 1165–1183.
19. Rosenfeld, A. and Smith, C., 1997. "Recent developments in radiocarbon and stylistic methods of dating rock-art." *Antiquity*, 71, 405–11.
20. Bednarik, R. G., 1995. "The Côa petroglyphs: An obituary to the stylistic dating of Paleolithic rock-art." *Antiquity*, 69, 877–83.
21. Watchman, A., 2000. "A review of the history of dating rock varnishes." *Earth Science Reviews*, 49, 261–277.
22. Reimer, P. J. et al., 2013. "Selection and treatment of data for radiocarbon calibration: An update to the International Calibration (INTCAL) Criteria." *Radiocarbon*, 55, 1923–1945.

23. Reimer, P. J. et al., 2013. "INTCAL13 and marine 13 radiocarbon age calibration curves 0–50,000 years cal BP." *Radiocarbon*, 55, 1869–1887.
24. D'Errico, F. et al., 2001. "Les possible relations entre l'art des caverns et la variabilité climatique rapide de la dernière période glaciare." in Barrandon, J-N. et al., eds., *Twenty-first Reoncontres Internationales d'archéologie et d'histoire d'Antibes, Éditions APDCA*, Antibes, 333–347.
25. Bradley, R. S., 1999. *Paleoclimatology*, Second edition, Elsevier Academic Press.
26. Higham, T.F.G., 2011. "European Middle and Upper Paleolithic radiocarbon dates are often older than they look: Problems with previous dates and some remedies." *Antiquity*, 85, 235–249.
27. Combier, J. and Jouve, G., 2012. "Chauvet's cave art is not Aurignacian: A new examination of the archaeological evidence and dating procedures." *Quatär*, 59, 131–152.
28. Cuzange, M-T. et al., 2007. "Radiocarbon intercomparison program for Chauvet cave." *Radiocarbon*, 49, 339–347.
29. Higham, T.F.G. et al., 2011. "Precision dating of the Paleolithic: A new radiocarbon chronology for the Abri Pataud (France), a key Aurignacian sequence." *J. Human Evolution*, 61, 549–563.
30. Thery-Parisot, I. and Thiebault, S., 2005. "Le Pin *(Pinus Sylvestris)* preference d'un taxon ou containte de l'environment? Etude des charbons de bois de la Grotte Chauvet." *Bull. Soc. Préhistorique Francaise*, 102, 69–75.
31. Sautuola, M. Sanz de, 1880. *Breves apuntes sobre algunos objetos prehistóricos de las provincial de Santander*. Santander: Telesforo Martinez (facsimile and translation, Turner 2004).
32. Lawson, A. J. 2012. *Painted Caves: Paleolithic Rock Art in Western Europe*. Oxford University Press.
33. Valladas, H. et al., 1992. "Direct radiocarbon dates for prehistoric paintings at the Altamira, El Castillo and Niaux caves." *Nature*, 357, 68–70.
34. Moure Romanillo, A., 1997. "New absolute dates for pigments in Cantabrian caves." *International Newsletter on Rock Art*, 18, 26–28.

7 The Cave Art Fallacy

As pointed out in chapter 1, a commonly held view among prehistorians of cave art is that the hunter-painters did not paint what they saw in their immediate environment; this view is based on the apparent fact that there is little correlation between kitchen bones (reflecting diet) and the animals depicted. Apparently, there were cultural filters that determined the animals that were portrayed, involving ritual or some type of symbolic representation (see chapters 1 and 10). These ideas extend back to the mid-nineteenth century, even before Paleolithic cave art was discovered, and are based on animal depictions on portable art and the mismatch with bone remains at certain Upper Paleolithic open air sites (Kesserloch, Switzerland, for example).

The Magdalenian open-air site of Gönnersdorf, on the Rhine River in Germany, indicates a mismatch between kitchen bones and animals on portable art—mammoths are commonly engraved on pieces of slate (sixty one in all), but mammoth bones are rare in kitchen refuse at the habitation site, whereas reindeer and horse dominate.[1] Conversely, reindeer depictions are rare on the portable art. Similarly, at La Vache, a cave near the French Pyrenees, ibex dominate the kitchen refuse, but horse and bison dominate the portable art.[2] Taken at face value these portable art examples indicate to prehistorians that the painters "did not paint what they ate."

This pattern was apparently confirmed by subsequent cave art discoveries.[3] Historically, the best-known example of the mismatch between excavated bones and cave imagery is Lascaux, where 88 percent of bones were reindeer, but the cave representations show only one or possibly a few reindeer out of hundreds of animals.[4] A second difficulty that is sometimes raised is that while aurochs are commonly depicted at Lascaux, aurochs bones are not found in the region—but this neglects the fact that aurochs bones are difficult to distinguish from bison, and both are usually labeled simply as bovines. A similar mismatch between cave bones and depicted cave art occurs at Ekain, in the Basque region of northern Spain.[5] This has led to the conclusion that the painters did not paint what they hunted or saw, and further, that the environment (e.g., climate) did not play a role in what they depicted.[6] It is possible, however, that they did depict what was in their environment, *except* what they hunted and ate. This would still result in a mismatch between cave depictions and kitchen bones.

The underlying assumption in all of this is that the refuse bones are the same age as the cave artwork, but this has *never* been demonstrated. There appears to be a pattern of accepting observations at face value on the part of paleohistorians, without critical evaluation. As outlined in the introduction, the Lascaux cave was poorly excavated, and much of the archaeological record was destroyed by workmen clearing the cave for tourists in the 1940s and 1950s.[7,8] As the French prehistorian Jean Clottes noted, "in most of the deep caves, archaeological evidence was destroyed as they were discovered." [9] At the Magdalenian cave site of Kesslerloch, Switzerland, excavated in 1873, more than a ton of bones were dug up from the archaeological layer, placed in wicker baskets, and washed in a nearby stream so that the bones would not be damaged—but at the same time, most of the archaeological stratigraphic information was lost.[10] Such destruction was mainly by well-meaning amateurs who dug artifacts "like potatoes." Moreover, the standards of archaeology in the nineteenth century were not up to the task of making the necessary fine spatial and stratigraphic distinctions required for detailed correlations. It is sad to contemplate that all this happened even before carbon-14 dating was developed in the 1950s.

At Gönnersdorf, Germany, ignoring for now the fact that mammoths were unlikely to be carried back to the encampment, and so are not expected to figure prominently in the excavated bone assemblage, a more serious issue concerns the relative ages of the bones and the portable art. Modern radiocarbon dating shows that the bones at Gönnersdorf span the range of 18 ka to 14 ka (2 σ) calibrated years.[11] However, dates on the mammoth bones and rhinoceros bones are older than the hunted fauna bones (bison, reindeer, and horse), and there appears to be a hiatus or gap between the two groups of ages.[11] A simple explanation of the apparent mismatch between bones (hunted fauna) and portable art (mainly mammoth) is that the mammoth depictions belong to the earlier (colder) occupation of the site. It is very unlikely that the mammoth images belong to the younger age range because woolly mammoths were extinct by 14 ka cal BP (see chapter 4). Therefore, the apparent mismatch at Gönnersdorf is likely due to comparison of bones and portable art of different ages.

The error associated with a single modern carbon-14 date (including laboratory analytic error, plus the error associated with the calibration curve) is typically ± 500 years, and usually much larger, when several dates are combined together. This error typically encompasses several climate regimes, so the probability of radiometric dates on excavated bones agreeing with directly dated cave imagery is low. The problem is more severe, considering that, in the past, many of these ages suffered from contamination with modern carbon and were inaccurate (see chapter 6). Moreover, until relatively recently, un-calibrated radiometric dates were commonly used by prehistorians, together with outmoded continental Ice Age chronologies, making any correlation between animals depicted and climate unlikely to be successful.[6]

The bones from archaeological sites reflect choices made by the hunters themselves, which depended on the season, climate, and terrain.[12] They would have preferentially focused on species that gave the greatest return for the least amount of effort. Moreover, for the larger animals, such as bison or mammoth, the hunters may have taken meat off the bones at the hunting site, rather than carry the carcass back to a permanent base camp, leaving no bone record of that animal. The Magdalenian site at Verberie near Paris appears to have been such a butchering site for reindeer (chapter 2). In addition, caves were commonly flooded during ice retreats, disrupting old strata and laying down new strata. The archaeological bone record therefore is not representative of the animals in the environment as a whole, and in many cases, the strata have been disturbed. But there is a more fundamental reason why bone remains commonly do not correspond to depicted faunal assemblages in cave imagery.

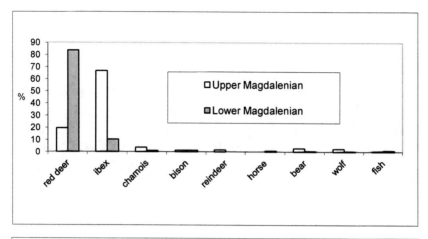

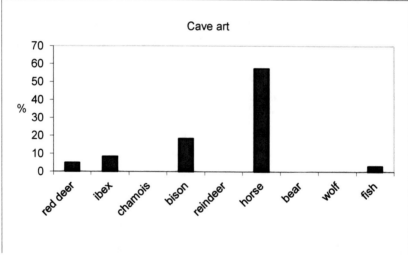

Figure 7.1 Upper panel: Distribution of bones from two different archaeological layers at Ekain cave, northern Spain. Red deer dominate in the lower layer, and ibex dominate in the upper layer. **Lower panel:** Animals depicted on the walls at Ekain. Horses and bison dominate, distinctly different from bone remains in either the upper or lower archaeological layers.

The case of Ekain, northern Spain[5], illustrates the problem. At Ekain, excavated layers, one of Lower Magdalenian and one of Upper Magdalenian age, yield different assemblages of bones. The presence of bone harpoons in the upper layer indicates an Upper Magdalenian age. This layer is dominated by ibex bones, followed by less-abundant red deer, whereas the lower layer is dominated by red deer, followed by less-abundant ibex (figure 7.1, upper panel). In contrast, the cave art is dominated by horse, followed by bison, ibex, and deer in decreasing order (figure 7.1, lower panel). An engraved bone bearing the image of a deer, an ibex, and a horse from a late Magdalenian layer was dated at 12.1 ± 190 ^{14}C BP, yielding a calibrated range of 13.4–14.5 ka cal BP (2σ), confirming the Upper Magdalenian stratigraphic age.[13]

Assume, for the sake of this argument, that the cave art was directly dated by the radiocarbon method and yielded a calibrated age range of 14–15.5 ka BP (2σ). In that case, the hypothetical cave art age overlaps with the bone age range. When these ranges are plotted on the graph of the GISP2 methane climate curve (chapter 3), an interesting feature, not previously recognized, emerges (figure 7.2, upper panel). The bone age range corresponds to a temperate-moist climate, consistent with the dominance of ibex and red deer in the Upper Magdalenian layer (based on the animal clusters identified in chapter 5). However, the cave art age range is consistent with all three climate types—temperate-moist climate; an intermediate climate; and a cold-dry climate. There is only one chance in three (33 percent) that the cave art age range and the bone assemblage will agree in terms of climate, even though the bone age and cave art ages overlap within error. In other words, we could say the bone age and the cave imagery are the same, but indicate different animal assemblages because they reflect different climates.

A similar situation exists at Niaux, where two bison have been directly dated, yielding a calibrated age range of 14.8–17.2 ka BP (2σ).[14] Across the river valley from the Niaux, another cave site, La Vache, was inhabited at about the same time and yields radiometric dates on bones with a similar calibrated age range, 14.2–17.0 ka BP, which overlaps the Niaux cave imagery dates.[15] Figure 7.2 (lower panel) plots these ages on the methane climate curve over the same time interval. It can be seen that the Niaux cave art dates are consistent with a temperate and a cold climate, but the La Vache bone dates are consistent with all three climates.

The depicted fauna at Niaux are dominated by bison, followed by horses and ibex. In contrast, the kitchen refuse at La Vache is dominated by ibex, followed by reindeer, with some red deer present.[15] There are only a few horse bones and no bison present. Although the age range of the two assemblages have substantial overlap, they allow for three different climate assemblages, and there is again only a 33 percent chance that the bone assemblage and the cave imagery agree with each other. The bone assemblage at La Vache also disagrees with the portable art imagery in that cave.[2] The strata at La Vache were poorly excavated but nevertheless indicate several

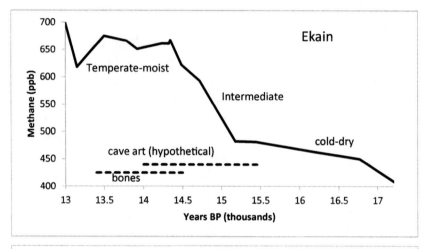

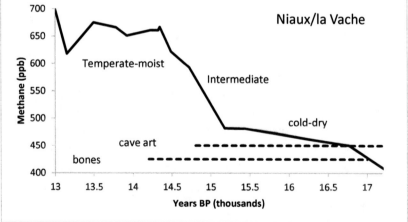

Figure 7.2 Upper panel: Ekain. Different climate regimes between 13,000 and 17,000 years ago. Also shown is a hypothetical age range for dated cave art at Ekain. This age range is consistent with three climate regimes—temperate, intermediate, and cold. **Lower panel:** Niaux/La Vache. The same climate regimes are shown as above. The directly dated cave art age range is shown and the age range associated with bones at the nearby La Vache cave is also shown. The cave bones are consistent with three different climates.

human occupation levels[2,15], allowing for the possibility that the bone assemblage and the Niaux cave imagery are of different age, even though they overlap in age within error. As pointed out in chapter 6, the painters of Niaux probably inhabited La Vache. The portable art depictions at La Vache are also similar to cave art at Niaux (horse, bison, ibex). The kitchen refuse (mainly ibex) reflects a different time frame and climate than that of Niaux.

The two examples above (Ekain and Niaux/LaVache) show that, even if the age range of the bone assemblage and the cave art overlap within error, during periods of rapid climate change, differences between the two assemblages can be expected, and in the case of these examples, a mismatch between bone assemblage and depicted cave art it is the most probable outcome. I refer to this as the "they did not paint what they ate fallacy" or simply the Cave Art Fallacy. The recognition of the Cave Art Fallacy is only possible because of three major scientific advances in the past few decades: 1) direct dating of cave imagery by accelerated mass spectrometry; 2) the calibration of ^{14}C dates back to 50,000 years ago; and 3) a precise climate record from polar ice cores.

The examples of Ekain and Niaux/La Vache (and Gönnersdorf) will no doubt be rejected for various reasons specific to those examples. It is important therefore to generalize the Cave Art Fallacy to the Upper Paleolithic in the time interval of 12–45 ka (figure 7.3). This figure is not based on any particular cave and so is independent of any specific archaeological framework. Figure 7.3 has three essential features: 1) a calibrated climate curve (ice core) with a half-wavelength of 1,000 years, typical of this time period that spans a cold and a temperate climate; 2) a calibrated direct date on the cave imagery; and 3) a calibrated date on presumably related bone material overlapping the cave art imagery within error (typically ± 300 years). With these conditions, overlapping bone and cave art imagery dates can represent different climate regimes and therefore can be expected to differ in their faunal assemblages. The Cave Art Fallacy should not apply to the Holocene (less than 11,000 years BP), when the climate stabilized. Indeed, the postglacial art works of eastern Spain (Levantine art) depict animals in broad agreement with their remains.[16]

Lascaux, one of the chief examples of "they did not paint what they ate", was poorly excavated, and the imagery cannot be directly dated. The charcoal ages from Lascaux span several thousand years[17], and it has even been suggested that the shaft, where four of the five radiometric dates come from, actually belongs to a different cave.[18] The animals depicted at Lascaux (aurochs, red deer, and horse) were likely present during the Lascaux interstadial at 19 ka BP. During the LGM, reindeer were likely present immediately before and after the Lascaux interstadial. A radiocarbon date on a reindeer antler from the shaft at Lascaux (18.6 ka ± 190 ^{14}C BP)[17] yields a calibrated age range of 22.0–22.9 ka cal BP (2σ)—too old for the Lascaux interstadial. In other words, whoever ate the meat of that reindeer did not paint at Lascaux. Lascaux, often upheld as a paradigm, is a poor example of "they did not paint what they ate" idea.

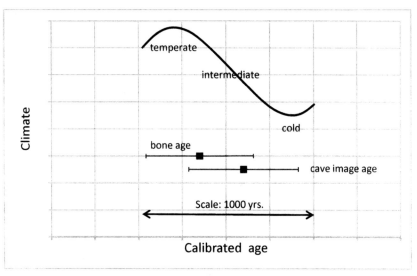

Figure 7.3: This figure is a generalization of figure 7.2 without reference to a specific cave. The diagram has three main features: a calibrated climate curve with a half-wavelength of 1,000 years allowing for three climate regimes within that time frame, and a radiocarbon bone date which overlaps with a cave art date with error bars of ± 300 years. The probability of the bones and cave art agreeing is only one in three. A mismatch between bone refuse and cave art is therefore likely, even when the dates overlap.

Two parallel studies by American anthropologists Rice and Paterson shed further light on the relationship between animal bones and animals displayed in cave imagery.[19,20] One study in southwest France, and the other in northern Spain, showed that when broken down by sub-region, there was good correlation between bones and cave art. When animal weight was included (reflecting meat yield) the correlation was substantially increased for both areas combined (R = 0.8). These authors aggregated the bone remains over time, thereby providing a time-averaged sample for each sub-region. There is good correlation between these time-averaged samples and cave imagery, indicating that the cave painters did not conjure up the animals from nowhere in some type of shamanic trance or ritual. Rather, their imagery reflects the real environment. These two studies strongly support the Cave Art Fallacy. The Cave Art Fallacy removes a major impediment to showing that cave imagery accurately reflects the climate at the time it was created. The next chapter tests this hypothesis using the GISP2 methane climate curve.

References Cited

1. Bosinski, G., 1973. "Le site Magdalenian de Gönnersdorf." *Bull. Soc. Préhistorique Ariège*, 28, 25–48.
2. Delporte, H., 1993." L'art mobilier de la grotte La Vache: Premier essai de vue générale." *Bull. Sociéte Préhistorique Francaise*, 90, 131–137.
3. Bahn, P. G. and Vertut, J., 1997. *Journey through the Ice Age*. University of California Press.
4. Bouchud, J., 1979. "La faune de la grotte de Lascaux." In *Lascaux Inconnu*, Leroi-Gourhan, Arl. and Allain, J. eds., 147–152, CNRS, Paris.
5. Altuna, J., 1983. "On the relationship between archaeo-faunas and parietal art in caves of the Cantabrian region." In *Animals and Archaeology: Hunters and their prey*, Clutton-Brock, J. and Grigson, C. eds., 1, 227–238. British Archaeological reports, International Series, 163, Oxford.
6. Alcolea, J. J. and de Balbín, R., 2003. "Témoins de froid. La faune dans l'art rupestre paléolithique de l'interior péninsulaire." *L'Anthropologie*, 107, 471–500.
7. Leroi-Gourhan, Arl., 1979. "La stratigraphie et les fouilles de la grotte de Lascaux." In *Lascaux Inconnu*, Leroi-Gourhan, Arl. and Allain, J. eds., 45–74, CNRS, Paris.
8. Delluc, B. and Delluc, G., 2006. *Discovering Lascaux*. (English edition), Éditions SudOuest.
9. Clottes, J. and Lewis-Williams, D., 1998. *The Shamans of Prehistory: Trance and magic in the painted caves*. Abrams, New York.
10. Merck, K., 1876. *Excavations at the Kesslerloch site near Thayngen, Switzerland*. Longmans, London.

11. Stevens, R. E. et al., 2009. "Radiocarbon and stable isotope investigations at the Central Rhineland sites of Gönnersdorf and Andernach-Martinsberg, Germany." *J. Human Evolution*, 57, 131–148.
12. Hervé, C., 1984. "Le Tableau de Chasse." *Histoire et Archeologie*, 87, 26–28.
13. Lawson, A. J., 2012. *Painted Caves: Paleolithic Rock Art in Western Europe.* Oxford University Press.
14. Valladas, H. et al., 1992. "Direct radiocarbon dates for prehistoric paintings at the Altamira, El Castillo, and Niaux caves." *Nature*, 357, 68–70.
15. Pailhaugue, N., 1995. "La faune de la salle Monique, grotte de La Vache (Ariège)." *Bull. Soc. Prehistorique Ariège-Pyrénées*, L, 225.
16. Boado, F. C. and Romero, R. P., 1993. "Art, time and thought: A formal study comparing Paleolithic and postglacial art." *World Archaeology*, 25, 187–203.
17. Valladas, H. et al., 2013. "Dating French and Spanish prehistoric decorated caves in their archaeological context." *Radiocarbon*, 55, 1422–143.
18. Bahn, P. G., 1994. "Lascaux, composition or accumulation." *Zephyrvs*, 47, 3–13.
19. Rice, P. C. and Paterson, A. L., 1985. "Cave art and bones: Exploring the interrelationships." *American Anthropologist*, 87, 94–100.
20. Rice, P. C. and Paterson, A. L., 1986. "Validating the cave-art archeofaunal relationship in Cantabrian Spain." *American Anthropologist*, new series, 88, 658–667.

8 Cave Art and the Climate Curve

The main conclusion from chapter 5 is that just two variables can explain 68 percent of the variation of animals depicted in thirty-two French and Spanish caves. The two variables were temperature and moisture, interpreted as representing Ice Age climate: A cold-dry climate during stadial periods and a temperate-moist climate during interstadials. The remaining 32 percent of the variance might be explained by any one of the six theories listed in chapter 1.

The cold-adapted group of animals identified in chapter 5 was reindeer, woolly mammoth, woolly rhino, giant elk, and bison. The temperate group consisted of ibex, horse, aurochs, and red deer. The Chauvet cave was given as an example of a cold faunal assemblage and Lascaux as an example of the temperate assemblage, although nearly all caves consist of a mix of both assemblages. In chapter 4, it was pointed out that the ibex favors steep and rocky terrain and that cave localities close to higher elevations have higher numbers of ibex depicted. Although they belong to the temperate group, they are regarded as a better indicator of relief and elevation than of climate type.

If the animals depicted accurately reflect the climate at the time, then the proportion of cold-adapted and temperate groups of animals should reflect the atmospheric methane concentration at that time. The animal assemblage represented in each cave can be plotted on the methane climate curve (see figure 3.1) if we know the age of the cave imagery and if we had a measure of methane concentration in the atmosphere at that time. On the methane climate curve, the methane concentration ranges from a high of about 600 ppb (parts per billion) to a low of about 400 ppb over the time range we are interested in here. A climate index variable can be defined as follows, allowing each cave to be plotted on the vertical axis of the climate curve:

Methane (ppb) = (% cold assemblage x A) + (% warm assemblage x B)

where A and B are constants with values of 4.0 and 6.0 respectively. A cave with a 100 percent temperate assemblage (e.g., Chimeneas) would have a maximum methane concentration of 600 ppb. Conversely, a cave with a 100 percent cold-adapted assemblage would have a methane concentration of 400 ppb. A continuum exists between these two end-members. The underlying assumption in this formulation is that the faunal assemblages depicted reflect the climate at that time. The recognition of the Cave Art Fallacy (Chapter 7) permits this assumption. In this chapter I test this climate assumption using the GISP2 methane climate curve.

The simple expression above for the methane concentration is linear; in other words, an increase in one of the percentages results in a proportional increase in methane concentration, and it is independent of time. As pointed out in chapter 3, however, the glacial-interglacial cycles follow a sawtooth pattern, and this pattern has been modeled using a rather complex series of negative exponential functions dependent upon time, going back 110,000 years in the case of the GISP2 record.[1] Such mathematical complexity is not warranted here, especially with such a small database of directly dated caves, and the reader should be aware that the above simple linear expression can only be an approximation. In the ice core record, the interstadials become smaller toward the end of a glacial cycle, and therefore, their amplitude depends on time. The model will prove wanting in two cases, namely Cosquer phase II and Chimeneas, which are dated to the last glacial cycle and occur during interstadials documented by pollen records (Laugerie and Lascaux interstadials, respectively)[2], but these interstadials are poorly represented on the methane curve. They are nevertheless prominent on the GISP2 oxygen isotope record.[3]

Table 8.1 shows the corresponding methane concentration for each directly dated cave using the simple linear model above. The percentages of cold-adapted and temperate-adapted fauna are from table A5.2 (chapter 5). The direct dates are calibrated using INTCAL[4] and are listed at the 2σ

confidence level. The dates are from animals depicted rather than from abstract signs.

Table 8.1. Parameters Used to Plot Directly Dated Caves on the Climate Curve

Cave	% Cold	% Temperate	Methane (ppb)	Age range (ka cal BP)
Chimeneas	0.0	100.0	600	17.9 - 18.6
La Pasiega	6.0	94.0	588	13.4 - 17.0
Cosquer II	9.3	90.7	581	21.3 - 24.2
Las Monedas	26.9	73.1	546	13.5 - 14.2
Altamira	30.7	69.2	538	15.2 - 18.6
Le Portal	37.0	63.0	526	13.1 - 14.7
Cougnac*	40.9	59.1	471	26.2 - 30.3
Niaux	52.9	47.1	494	14.8 - 17.2
Pech Merle	55.6	44.4	489	27.8 - 29.5
Chauvet	69.3	30.7	461	35.4 - 36.3

*Ibex excluded

For cave artwork that has been deemed stylistically inhomogeneous, possibly involving painting over an extended period of time, the climate index would represent an average climate over that time period. In the case of the Cosquer cave (Bouches-du-Rhône, France), direct dating suggests two periods of decoration widely separated in time, with most of the animals painted during phase II.[5] This makes it difficult to plot Cosquer I on the climate curve. On a shorter time scale, it is assumed that the cave imagery reflects the animals encountered by the hunters on a seasonal basis whether within a stadial or an interstadial period. In other words, animal assemblages were largely determined on a seasonal basis.

The caves shown in table 8.1 have direct radiometric dates and, in some cases, stylistic dating does not agree with the radiometric dating. Tito Bustillo, in Asturias, northern Spain, is not included because its radiometric dates are discordant and appear to be contaminated with modern carbon.[6] El Castillo, in Santander, Spain is also not included, as it appears to have had a complex history of decoration.[6] Other caves with direct dates are not included because their animal inventories were not available to the author. Each cave is briefly discussed below. Their location is indicated at the end of the book.

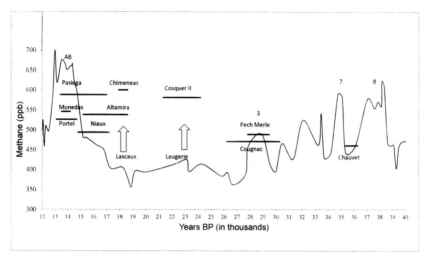

Figure 8-1: Plot of the methane climate curve from GISP2 together with ten cave imagery dates. Also shown are two pollen interstadials: Lascaux and Laugerie. Within error, eight of the ten caves fall on the curve. Chimeneas and Cosquer II coincide with the pollen interstadials. AB- Allerød-Bølling; numbers indicate Greenland interstadials.

Chimeneas *(Santander, Spain)*: Discovered in 1953, this cave is located in Santander, northern Spain. It is the only cave on the list in table 8.1 with 100 percent of the faunal assemblage belonging to the temperate group, and it plots highest on the methane scale at 600 ppb. Leroi-Gourhan assigned it to his style III, corresponding to the late Solutrean and early Magdalenian period.[7] The Spanish archaeologist Gonzalez Echegaray assigned the cave artwork to the Lascaux interstadial[8], which in Spain belongs to the Solutrean. This interstadial is based on pollen data (discussed in chapter 6), and it occurs at about 19 ka ± 500 BP in calibrated years. A direct radiocarbon date on a deer[9] gives an age of 15.1 ka ± 140 ^{14}C BP equal to 18.6–17.9 ka cal BP. This age range is consistent with the age of the Lascaux interstadial (see figure 8.1). As noted already, this interstadial does not show on the methane ice core record because it occurs late in the glacial cycle.

Las Monedas *(Santander, Spain)*: Also discovered in 1953, this cave is geographically very close to Chimeneas. Leroi-Gourhan confidently assigned this artwork to his youngest style, IV, corresponding to the Middle to Late Magdalenian. It also plots high on the methane curve (68.2 percent warm-adapted, 31.8 percent cold-adapted) at 546 ppb. Two dates on an ibex give 12.2 ka ± 110 BP and 11.6 ka ± 120 BP, and a direct age on a horse gives 11.9 ka ± 120 BP[9], corresponding to a calibrated age range of 13.5–14.2 ka BP. This places the cave close to the climate curve before the Bølling warm period (see figure 8.1). Some time ago, in 1974, Gonzalez Echegaray used Las Monedas and Chimeneas as examples of cave artwork undertaken under contrasting climate conditions, relatively cold and temperate, respectively.[8] Using the climate index proposed here, however, they both have substantial temperate faunal representations (Las Monedas 73 percent and Chimeneas 100 percent).

La Pasiega *(Santander, Spain)*: This cave, discovered in 1911, is part of a group of painted caves on a hill, Monte Castillo, including Las Monedas and Chimeneas. Excavations indicate Solutrean and Early Magdalenian deposits near one of the entrances, and Leroi-Gourhan assigned the paintings to this same time period, his style III. There appear to be two phases present: an earlier red paint phase and a later black paint phase with engravings. The cave is dominated by a temperate assemblage (93.6 percent) corresponding to a methane concentration of 588 ppb, similar to the nearby but slightly older Chimeneas cave. A black engraved ibex yields a direct radiocarbon date of 13.7 ka ± 130, and a black painted bison yields two dates of 12.5 ka ± 160 and 12.0 ka ± 170[5], corresponding to a calibrated range of 17.0–16.2 ka and 15.2–13.4 ka, respectively, spanning a total of 4.6 thousand years for the three dates. These ages correspond to the Middle to Late Magdalenian, making them somewhat younger than the stylistic assignment above (Solutrean to Lower Magdalenian). La Pasiega lies high on the climate curve close to the Bølling warm peak. The cave is dominated by red deer (chapter 5).

Le Portel *(Ariège, France)*: This cave was discovered in 1908 and is situated about 13 km northwest of the town of Foix in southern France, not far from the Pyrenees. Industrial bone and lithic artifacts, including a bone harpoon, indicate a Magdalenian occupation. Leroi-Gourhan recognized two phases of artwork, an earlier poorly preserved phase and a later phase. He dated the later phase to the late Magdalenian, his style IV, similar in age to Las Monedas. The assemblage of younger phase animals portrayed is 61.5 percent temperate, 38.5 percent cold-adapted, corresponding to 526 ppb methane using the formula above. Direct dates on two different horses[5] yield 12.2 ka ± 125 ^{14}C BP and 11.6 ka ± 150 ^{14}C BP, corresponding to a calibrated age range of 13.1–14.7 ka cal BP. This cave plots very close to Las Monedas in the Allerød-Bølling time period.

Cougnac *(Lot, France)*: Discovered around 1950, it is located just north of the town of Cahors in south central France, about 100 km (62 mi) west of the Massif Central. Of the total of twenty-two animals depicted, the breakdown is as follows: eight ibex, four red deer, six mammoths, one horse, and three giant elk, which is an unusual assemblage, especially considering the mammoths and red deer together and that bison and aurochs are missing. The large number of ibex may reflect proximity to the Massif Central, as they would have been forced to lower elevation during cold periods. The high percentage of ibex (36 percent) distorts the climate index, so that ibex are omitted for this cave, giving the cold-adapted and temperate assemblage percentages of 64.3 percent and 35.7 percent respectively, corresponding to a methane concentration of 471 ppb. Although Leroi-Gourhan noted the atypical faunal assemblage, he assigned the artwork to a single homogeneous period corresponding to the late Solutrean and early Magdalenian, his style III. Four direct dates on two of the giant elk figures give: 25.1 ka ± 390 and 19.5 ka ± 270 (for the female elk) and 23.6 ka ± 350 and 22.8 ka ± 390 (male elk).[5] One of the dates on the female elk is over 5,000 years younger than the other, which might be explained by later retouching of the figure, as some retouching was observed on some other figures. Omitting the youngest date, the calibrated age range is 26.2 ka–30.3 ka, corresponding to the Gravettian cultural period, which is substantially older than the stylistic estimate. Cougnac plots on Greenland interstadial three, before the Last Glacial Maximum. The animals depicted clearly show a mixed assemblage (mammoth and red deer), possibly reflecting rapidly changing conditions.

Pech Merle *(Lot, France)*: Discovered in 1922, this cave is close to Cougnac and has similarities in term of signs and themes.[7] It contains twenty-eight mammoths, thirteen bison, and one giant elk belonging to the cold-adapted group, and twelve horses, seven aurochs, and four deer representing the temperate group of animals. The climate index yields 55.6 percent cold-adapted and 44.4 percent temperate faunal assemblage, corresponding to 489 ppb methane. Two phases of painting have been proposed, so the climate index is a time-averaged climate over the course of the paining. The early phase includes the famous spotted horses, one of which yields a direct date[10] of 24.6 ka ± 390 ^{14}C BP. This age corresponds to a range of 27.8–29.5 ka in calibrated years BP and plots on the climate curve at Greenland interstadial three during the Gravettian. The second phase of painting is undated and is dominated by a mammoth/bison theme. This phase possibly belongs to the Last Glacial Maximum. Leroi-Gourhan attributed this cave to his style III (late Solutrean-early Magdalenian). With only one direct date, however, this cave is poorly dated but it does plot close to Cougnac, to which it is similar on stylistic grounds and in animal assemblage.

The spots on the famous horses at Pech Merle, illustrated in most cave art books, raise an interesting question: are these polka-dot horses fanciful or mythical creatures that have a symbolic meaning? Terms such as "sorcerer" and "unicorn" have been long applied to certain unusual cave art depictions. A recent genetic study of samples of Pleistocene horses found that, out of thirty-one samples, six contained the gene referred to as "leopard spotting," which would have produced horses that looked very similar to the Pech Merle spotted horses. So the painters probably accurately painted what they saw.[11]

Niaux *(Ariège, France)*: The paint recipes at this cave and the associated cave, La Vache, were briefly discussed in chapter 6. Situated in the French Pyrenees, this cave has been known for a long time but its importance was recognized in 1906, and it belongs to Breuil's group of six "giants". The most famous part of the cave is the Salon Noir, which contains 83 percent of the animals depicted. The climate index yields 62.1 percent cold-adapted and 37.9 percent temperate faunal assemblage, corresponding to 445 ppb methane. Leroi-Gourhan assigned the artwork to his style IV, or the Middle to Upper Magdalenian. A small bison in the Salon Noir gives a direct date of 13.9 ka ± 150 ^{14}C BP, and a charcoal line between an ibex and a bison gives a direct date of 13.1 ka ± 200 ^{14}C BP and a large bison nearby gives a direct date of 12.9 ka ± 160 ^{14}C BP.[12,13] The calibrated age range for these dates is 14.8–17.2 ka cal BP. Niaux plots on the climate curve immediately after the Last Glacial Maximum, just as the climate began to ameliorate. It is dominated by bison.

Altamira *(Santander, Spain)*: This cave was discussed in the chapter 6. The percentages of cold-adapted and temperate assemblages are 30.7 percent and 69.2 percent respectively, corresponding to 538 ppb methane (see table 8.1). Its stylistic age (Leroi-Gourhan's style III) and radiometric age range (15.2–18.6 cal BP) agree with each other, and this cave plots on the climate curve just above Niaux on a warming trend at the end of the Last Glacial Maximum, very close to Las Monedas and Le Portel.

Cosquer *(Bouches-du-Rhône, France)*: Discovered in 1991 by Henri Cosquer, a deep-sea diver, it can only be accessed today underwater on the Mediterranean coast near Marseille, in southern France. During glacial periods when sea level was lower, it would have been about 4 km to 5 km from the shoreline.[14] The cave artwork consists of a large number of negative hand stencils done with red paint. They are earlier than the animal depictions, which are mostly engravings, but about 20 percent are drawn in black paint. Two phases of artwork are recognized, but it was originally thought that the signs and hand stencils belonged to phase I and the animal depictions belonged to the later phase II, based on some direct AMS ^{14}C dates and some dates on charcoal from the floor. Additional direct dating later[15], however, showed that some of the animals depicted also belonged to the early phase, but most animal depictions belong to the later phase II.[5] The large herbivores depicted are dominated by horses (36 percent), followed by ibex (28 percent), red deer (17 percent), and bison (10 percent). Two giant elk are also listed. This is a temperate dominant assemblage (88.1 percent) with 11.9 percent cold-adapted, represented by bison and giant elk. Since most animals were depicted during the younger phase II period, this climate phase is assumed to have been temperate.

Direct dates on phase II animals (horses, bison, giant elk, and stag) range from 18.0 ka ± 190 ^{14}C BP for a bison to 19.7 ka ± 210 ^{14}C BP on a horse—corresponding to a calibrated age range of 21.3–24.2 ka cal BP. Phase II falls within the Solutrean cultural period. Two direct dates on a bison belonging to phase I yield ages of 27.4 ka ± 430 ^{14}C BP and 26.3 ka ± 350 ^{14}C BP, corresponding to a calibrated age range of 29.6–32.3 ka cal BP, corresponding to the Gravettian cultural period. It seems unlikely that the same animal assemblage used for phase II also applies to phase I, which occurred at least 5,000 years earlier. Stylistic considerations may be able to identify the animal assemblages belonging to these separate phases. Until then, Cosquer phase I cannot be plotted on the climate curve.

Phase II of Cosquer does not plot on the methane climate curve. It does, however, coincide with the Laugerie pollen interstadial dated at 19.2 ka ^{14}C BP (22.6–23.6 ka cal BP).[2] Site 976 of the Ocean Drilling Program from the Mediterranean Sea near the Gibraltar strait shows a spike in temperate tree pollen corresponding to interstadial two on the Greenland ice core.[16] Interstadial two corresponds to the Laugerie interstadial[2], but it is poorly represented on the methane curve. (figure 8.1).

Chauvet *(Ardèche, France)*: The controversy over the age of the Chauvet cave imagery was discussed in chapters 2 and 6. The essential disagreement is that the sophistication of the imagery is more consistent with a Magdalenian (or possibly Solutrean) cultural time frame, whereas direct dates of several drawings done in charcoal together with over 100 dates from ten different laboratories on several pieces of charcoal from nearby hearths yield an Aurignacian age (32,000 ± 200 ^{14}C BP), corresponding to 35.4–36.3 ka calibrated years before present. Direct dates on charcoal torch marks that overlie the paintings yield younger ages of about 26.0–27.0 ± 400 ^{14}C BP, corresponding to 29.3–31.6 ka cal BP—these dates apparently date a later visit to the cave but do not date any of the paintings themselves. The Chauvet faunal assemblage is 69.3 percent cold-adapted and 30.7 percent warm adapted and it plots on the methane curve between Greenland interstadials seven and eight (figure 8-1).

In summary, the pattern that emerges is that the caves with direct dates plot on or close to the climate curve. Five caves, Altamira, Niaux, Le Portel, La Pasiega, and Las Monedas, cluster around the warming period after the Last Glacial Maximum and before the Bølling maximum warm period at about 14,700 years ago. Two of these caves fall directly on the climate curve (La Pasiega and Niaux), and three fall slightly off. The error associated with the GISP2 timescale is 5 percent at 15,000 years BP, corresponding to ± 750 years.[17] Within this error, all five of these caves fall on the climate curve. Chauvet, Pech Merle, and Cougnac also fall on the climate curve. The two caves (Chimeneas and Cosquer II) that plot away from the curve are attributed to interstadials based on pollen analysis (Lascaux and Laugerie interstadials, see chapter 5). These caves therefore support the climate hypothesis and are regarded as consistent with the climate record.

To better determine whether this pattern could be due to chance alone, computer trials were performed in which the ages and methane values were assigned random values. Random ages between 12,000 and 45,000 years ago, with random methane concentrations between 400 and 600 ppb, were assigned to ten hypothetical caves. These random trials were plotted on the same diagram as figure 8.1, and the caves that fell on the curve within the error of the analysis were counted. An error of ±5 percent was associated with each data point on the X and Y axis. A ±5 percent error on 20,000 years corresponds to ±1,000 years, which is a reasonable error bar in the case of calibrated radiometric dates when taken together. A ±5 percent error on 500 ppb methane concentration corresponds to 25 ppb; this error could result from errors associated with the animal inventory for each cave. The process was repeated 100 times.

The average number of caves that fall on the climate curve in the random trials is five out of ten. In addition to random trials, Poisson statistics can also shed light on this issue. Poisson statistics are commonly used for a small number of discrete events when the average number of such events is known. Poisson statistics can be used here to evaluate the probability whether a cave falls on or off the climate curve. For an average value of five caves falling on the curve, the value observed in the random trials, the Poisson probability is 17.5 percent for this event. For one cave to fall on the curve the chance is 3 percent, and for all ten caves to fall on the curve the probability is 1.8 percent. The 100 random computer trials agree very well with these Poisson statistics.

The probability that the results presented here, namely eight out of ten caves falling on the climate curve, is 6.5 percent. If Chimeneas and Cosquer II are included, the probability that all ten caves are consistent with the climate record is 1.8 percent. From these small percentages (1.8 – 6.5 percent), it is concluded that climate played a major role in what was depicted in Upper Paleolithic Franco-Cantabrian cave imagery. The number of caves in this admittedly small sample is not likely to increase substantially in the future, due to the paucity of carbon-based pigments in this cave art. I conclude that climate was the dominant factor in determining the large herbivore faunal assemblages depicted in Franco-Cantabrian cave imagery from the Aurignacian to the Magdalenian. In the next chapter, a climate-based hypothesis is proposed for the Rouffignac and Chauvet woolly mammoth imagery.

References cited

1. O'Hara, K. D., 2008. "A model for late Quaternary methane ice core signals: Wetlands versus a shallow marine source." *Geophysical Research Letters*, 35, L02712.
2. Leroi-Gourhan, Arl., 1997. "Chauds et Froids de 60,000 a 15,000 BP." *Bull. Soc. Préhistorique Francaise*, 94, 151–160.
3. The Greenland Summit Ice Cores CD-ROM. 2003. National Snow and Ice Data Center and National Oceanographic and Atmospheric Administration, Boulder, CO.
4. Reimer, P. J. et al., 2013. "INTCAL13 and marine 13 radiocarbon age calibration curves 0–50,000 years cal BP." *Radiocarbon*, 55, 1869–1887.
5. Lawson, A. J., 2012. *Painted Caves: Paleolithic Rock Art in Western Europe*. Oxford University Press.
6. Fortea Perez, F. J., 2002. "Trente-neuf dates C14-SMA pour l'art pariétal paléolithique des Asturies." *Bull. Soc. Préhistorique Ariège-Pyrénées*, 57, 7–28.
7. Leroi-Gourhan, A., 1967. *Treasures of Prehistoric Art*. Abrams, New York.
8. Gonzalez Echegaray, J., 1974. "Pinturas y grabados de la cueva de Las Chimeneas (Puente Viesgo, Santander)." *Monografias de Arte Rupertre, Arte Paleolítico* No. 2, Barcelona.
9. Moure Romanillo, A. et al., 1997. "New absolute dates for pigments in Cantabrian caves." *International Newsletter on Rock Art*, 18, 26–28.
10. Lorblanchet, M. et al., 1995. "Direct date for one of the Pech Merle spotted horses." *International Newsletter on Rock Art*, 12, 2–3.
11. Pruvost, M. et al., 2011. "Genotypes of predomestic horses match phenotypes painted in Paleolithic works of art." *Proceedings National Academy Sciences*, 108, 18,626–18,630.
12. Clottes, J. et al., 1992. "Des dates pour Niaux et Gargas." *Bull. Soc. Préhistorique Francaise*, 89, 270–274.
13. Valladas, H. et al., 1992. "Direct radiocarbon dates for prehistoric paintings at the Altamira, El Castillo, and Niaux caves." *Nature*, 357, 68–70.
14. Clottes, J. and Courtin, J., 1996. *The Cave Beneath the Sea: Paleolithic Images at Cosquer*. Abrams, New York.
15. Clottes, J. et al., 1996. "New direct dates for the Cosquer cave." *International Newsletter on Rock Art*, 15, 2–4.
16. Nebout, C. N. et al. 2002. "Enhanced aridity and atmospheric high-pressure stability over the western Mediterranean during the North Atlantic cold events of the past 50 ky." *Geology*, 30, 863–866.
17. Brook, E. J. et al., 1996. "Rapid variations in atmospheric methane concentration during the past 110,000 years." *Science*, 273, 1087–1090.

Mammoth Migrations

A review of the woolly mammoth's role both in both cave and portable art suggests that this enormous beast, weighing over five metric tons, was an icon in the mind of the Upper Paleozoic hunter.[1] The mammoth fossil record from Aurignacian time to the end of the Last Glacial Maximum (40 ka–12 ka) covers approximately half the land area of the entire northern hemisphere.[1,2] The Rouffignac cave, which sparked the idea for this book (see preface), together with Chauvet are the two caves with the greatest number of depicted mammoths (table 9.1). At Chauvet, the mammoth appears to have been special enough to deserve its own symbol namely, ω (a lower-case omega, apparently representing a pair of mammoth tusks; seventeen examples in all), and these are called Chauvet-type signs.[3] No other animal in cave art has been assigned a specific symbol as far as is known.

Rouffignac is particularly enigmatic because woolly mammoths dominate the imagery, a total of 158 depictions representing 70 percent of all the animals. The Rouffignac faunal assemblage is 87.6 percent cold-adapted and 12.4 percent temperate, corresponding to a low methane concentration of 425 ppb. This methane concentration would plot on the climate curve (see chapter 8) at six different locations, so this climate index alone is not useful in constraining the age of Rouffignac. Leroi-Gourhan assigned it to his style IV, placing it in the middle to late Magdalenian.

The climate hypothesis developed in this book appears to apply particularly well to the Rouffignac and Chauvet caves. The question remains why depict 158 mammoths, when one would have sufficed to indicate their presence in the environment? It is possible that the cave painters had a different reason for devoting over half the cave to this animal (other animals depicted at Rouffignac include the woolly rhinoceros, bison, ibex, and horse). A brief look at the population dynamics of both humans and mammoths over the past 40,000 years in Europe provides some insight into this question.

Table 9.1. Mammoth Caves

Cave	Mammoths (%)	Age*+#	Department
Rouffignac	158 (70)	Middle Mag.#	Dordogne
Pair-non-Pair	2 (8)	Gravettian#	Gironde
Combarelles	7 (12)	Middle Mag.#	Dordogne
Gargas	4 (6)	Gravettian#	Hautes-Pyrenees
Les Trois Freres	2 (10)	Middle Mag.#	Ariége
Font-de-Gaume	23 (21)	Middle Mag#	Dordogne
Pech Merle	7 (10)	Gravettian*	Lot
Cougnac	6 (27)	Gravettian*	Lot
Chauvet	68 (29)	Aurignacian*	Ardeche
Chabot	16 (76)	Solutrean+	Gard
Ebbbou	1 (2)	Solutrean#	Ardèche
Oulen	5 (63)	Unknown	Ardèche
La Baume L.	7 (88)	Aurignacian+	Gard
Altamira	1 (1)	Middle Mag.*	Santander
Grottte du Cheval	8 (53)	Aurignacian#	Yonne
Mayenne-Sciences	2 (13)	Gravettian*	Quercy

Note: * direct age; + indirect age; # stylistic age

An analysis of human population dynamics during the Upper Paleolithic of Western Europe, based on the number of archaeological sites shows several interesting patterns.[4,5] Foremost is that there existed a human population refuge located in the Aquitane basin of southwest France, from Aurignacian to Magdalenian times, which was maintained even through the Last Glacial Maximum. The location of this refuge was largely determined by geography, and it includes the Dordogne region. With the Atlantic Ocean to the west, the Massif Central to the east, and the Pyrenees to the south, this corridor represented a kind of cul-de-sac for populations (both human and animal) during southward retreat from advancing ice sheets. A large population center also existed further south, in northern coastal Spain, especially during the Solutrean cultural period.

Human populations followed the animal populations, which in turn followed the changing vegetation. During cold periods, the polar desert and tundra-steppe vegetation migrated south, and during temperate periods, the tundra-steppe migrated north into what was previously polar desert. Human populations moved north into France and northern Europe after the Last Glacial Maximum, from about 15.5 ka cal BP onward.[4,5] The human population is estimated to have been 4,000 in the Aurignacian period, increasing to 8,000 at the Last Glacial Maximum, and then exploding to 40,000 by the end of the Magdalenian, when the ice had retreated and the northern European plains were recolonized.[4]

As outlined in chapter 4, two southward migrations of mammoths into Spain are recognized; one took place in the upper Aurignacian cold period (between interstadials seven and eight), and a second one during the Last Glacial Maximum.[6] The second migration into Spain is documented by mammoth finds at two northern Spanish sites at 23.6 ka and 21.8 ka cal BP.[6] The record of woolly mammoth finds to the north in Europe shows a distinct gap over the interval 24–19 ka in calibrated years BP[7], and this is illustrated in figure 9.1. Iberian (Portugal and Spain) mammoth dates are also shown on this diagram.[6] This gap during the Last Glacial Maximum likely reflects the southward migration at about 24 ka cal BP, and their northward return at about 19 ka, when conditions further north became tolerable for this already cold-adapted species. A similar gap exists for giant elk bones suggesting the gaps are real and not an artifact of the fossil record.[7] That these data represent migrations is indicated by the fauna associated with the mammoth finds in Spain—they are associated with temperate faunal assemblages such as red deer, a mixed assemblage not seen in European assemblages to the north.[6] The refuge areas of the Dordogne and northern Spain remained populated with humans throughout this time.[4,5]

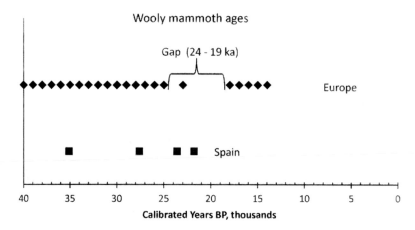

Figure 9.1: The woolly mammoth bone record in Europe excluding Iberia (Portugal and Spain). Note the gap in the data at about 24–19 ka cal BP. The gap is interpreted as due to the southward migration of the mammoth into Spain during the Last Glacial Maximum. Dates for woolly mammoths in Iberia are also shown (lower data points). An earlier migration at about 35 ka cal BP is suggested by mammoth finds in Spain at this time.

The hypothesis proposed here is simple: that Rouffignac and other caves with large numbers of mammoths depicted represent the documentation (and perhaps celebration) of migrating herds of mammoths south into Spain during stadial periods. The mammoths of the Chauvet cave are thought to reflect the Aurignacian migration of mammoths into Spain along an eastern route through the Rhône valley at 35-36 ka ± 300 cal BP, which is the calibrated radiometric age of Chauvet. Mammoth bone finds in southern Spain date to the interval 40.4–30.6 ka cal BP[8], a time frame that includes the Chauvet imagery. The nearby Chabot cave (see table 9.1) may also represent a Solutrean migration by the same eastern route through the Rhône valley during the Last Glacial Maximum. La Baume Latrone and Ebbou caves, also rich in mammoth depictions, lie along the same route. The hypothetical mammoth routes, two western routes and one eastern, are shown in figure 9.2. The coastline during the Last Glacial Maximum (LGM) is also shown when sea level was 100 meters lower, giving easier access to Spain along the coastline[9]. The mammoth routes are largely along major river systems which during stadials would have been low in water or dry.

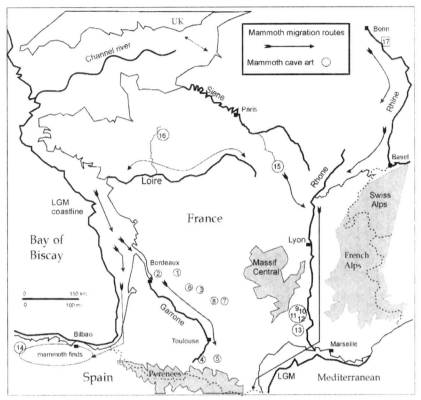

Figure 9.2: Map of southern France and northern Spain indicating likely mammoth migration routes into Spain. One of the western routes passes the Rouffignac cave and follows the Garrone River. The eastern route passes Chauvet along the Rhône valley. This route is strongly constrained by geography. The extended coastline during the Last Glacial Maximum is indicated. The caves depicting mammoths along these routes are indicated: 1. Rouffignac, 2. Pair-non-Pair, 3. Combarelles, 4. Gargas, 5. Les Trois Freres, 6. Font-de-Gaume, 7. Pech Merle, 8.Cougnac, 9. Chauvet, 10. Chabot, 11. Ebbou, 12. Oulen, 13. La Baume Latrone, 14. Altamira, 15. Grotte du Cheval, 16. Mayenne-Sciences, 17. Gönnersdorf (open air site).

Herds of migrating modern African elephants, up to a thousand strong, have been described, and mammoths probably displayed similar migration patterns in times of food shortage.[10] Search for water during dry stadials would have been as important as food. Such migrations would have taken several years, by analogy with modern elephant migrations, and must have been a spectacular sight. Such migrations would have been a literal journey through the ice age. If correct, Rouffignac would have been decorated about 24,000 years ago (late Solutrean, southward journey) or 19,000 years ago (early Magdalenian, return journey). The transient nature and rarity of these migrations accounts for the paucity of mammoth bones along these routes. Within the Rhône valley, however, worked mammoth remains found in loess deposits are testament to human exploitation of mammoth resources during a cold, dry episode, thought to be Gravettian in age.[11] In the next and final chapter, existing theories of cave art are briefly reviewed, and a general theory is proposed for Upper Paleolithic cave imagery, which explains why it lasted so long.

References cited

1. Braun, I. M. and Palombo, M. R., 2012. "*Mammuthus primigenius* in the cave and portable art: An overview with a short account of the elephant fossil record in southeastern Europe during the last glacial." *Quaternary International*, 276–277, 61–76.
2. Markova, A. K. et al., 2010. "New data on the dynamics of the *mammuthus primigenius* distribution in Europe in the second half of the late Pleistocene-Holocene." *Doklady Akademii Nauk*, 431, 547–550.
3. Azéma, M. and Clottes, J., 2008. "Chauvet-type signs." *INORA*, 50, 2–7.
4. Bocquet-Appel, J-P., 2000. "Population kinetics in the Upper Paleolithic in Western Europe." *J. Archaeological Science*, 27, 551–570.
5. Gamble, C. et al., 2004. "Climate change and evolving human diversity in Europe during the last glacial." *Phil. Trans. R. Soc. Lond.* B 359, 243–254.
6. Alvarez-Lao, D. J. and Garcia, N., 2012. "Comparative revision of the Iberian mammoth (*Mammuthus primigenius*) record into a European context." *Quaternary Reviews*, 32, 64–74.
7. Stuart, A. J. et al., 2004. "Pleistocene to Holocene extinction dynamics in giant deer and woolly mammoth." *Nature*, 431, 684–687.
8. Alvarez-Lao, D. J. et al., 2009. "The Padul mammoth finds-on the southernmost record of *Mammuthus Primigenius* in Europe and its southern spread during the Late Pleistocene." *Paleogeography, Paleoclimatology, Paleoecology*, 278, 57–70.
9. Lambeck, K. and Chappell, J. 2001. "Sea level change through the last glacial cycle." *Science*, 292, 679–686.
10. Lister, A. and Bahn, P. 1994. *Mammoths*. Macmillan, USA.
11. Gély, B., 2005. "La Grotte Chauvet a Vallon-Pont-d'Arc (Ardèche): Le context régional paléolithique". *Bull. Soc. Préhistorique Francaise*, 102, 17–33.

10 Cave Imagery as Cultural Meme

The strongest argument in favor of a climate control on cave imagery is that it lasted approximately 25,000 years with a coherent thematic unity over that time. When the climate stabilized in the Holocene epoch, this type of cave imagery disappeared. Climate change was the metronome that paced European cave imagery from Chauvet (36, 000 years BP) to Le Portel (13,000 years BP). In other words, the species depicted changed according to the climate over time. Political, philosophical, cultural, or religious traditions have never lasted so long.

Some of the more significant theories of cave art were briefly mentioned in chapter 1.[1,2,3,4] To put the climate change theory proposed here in context, some of these theories are briefly reviewed before proposing a more detailed explanation of why this cave imagery lasted so long. A hierarchical diagram of some of these ideas is presented in figure 10.1.

One of the earliest theories was that of "art for art's sake," originally proposed for portable art, whereby life was assumed to be easy for the Paleolithic hunters, who had time on their hands. As pointed out in chapter 3, during stadial periods, survival was probably quite difficult. As cave imagery discoveries became more common, it appeared that much of the cave art was in inaccessible parts of the caves, not a likely location if cave painters were just demonstrating their artistic skill—this largely led to the demise of this theory. However, much of cave imagery could originally have been either near the cave entrance or in the open air, but because of exposure to the elements, few of these are preserved.[5] A revival of this theory was attempted in more recent years, but many of the comments on that paper appear to be skeptical.[6]

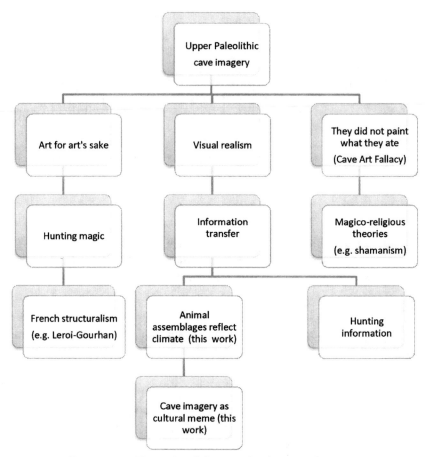

Figure 10.1: Hierarchical diagram for theories of cave art.

Henri Breuil himself favored the "hunting magic" hypothesis, in which the act of portraying the animal in a drawing increased the success of the hunt, a hypothesis originally proposed in 1903 by S. Reinach.[7] Speared bison make up about 15 percent of the cave images, which is a large percentage when you consider that only rarely would a hunter take a shot at an animal, and only if he could be confident of bringing the animal down.[8] The hunting magic theory was popular until the 1950s.

Very influential among the French is André Leroi-Gourhan and

Laming-Emperaire's theory[9], which grew from the French intellectual movement of structuralism popular at that time, and it involved abstract signs and animal groupings in the caves along symbolic gender lines. Leroi-Gourhan's contribution to the documentation of Franco-Cantabrian cave imagery is still important—especially the spatial distribution of different species within the caves—but his semiotic analysis of cave imagery has not stood the test of time.[1,2] In figure 10.1, this theory is represented on the left hand side, below hunting magic, but no evolutionary connection between the two is implied.

The conclusion that the painters "did not paint what they ate," which was reached very early on in the nineteenth century is still widely accepted—this is the Cave Art Fallacy of chapter 7. The assumption has been that the excavated bones (reflecting diet) are the same age as the depicted cave images, but this has not been demonstrated in most cases. This fallacy opened the way for a variety of theories that are grouped as magico-religious hypotheses, and which include totemism, fertility magic, and shamanism. Shamanism, which involves hallucinations or a trance state on the part of the shaman or medicine man, was applied to European cave art some time ago[10] and was popularized more recently with a well illustrated book[4]. An example of a shamanic hallucination in Paleolithic cave art is the polka-dotted horses at Pech Merle.[10] In chapter 8, however, it was pointed out that this horse likely had a gene variation termed "leopard complex spotting," which was not uncommon in Pleistocene time, and the hunter-painter probably actually saw this unusual spotted horse in the field. The unusual hide may have been highly prized. Some paleohistorians of cave art do not appreciate this intrusion of hard science into their discipline.

Viable theories of European cave art are shown in the central portion of figure 10.1. These theories favor realistic portrayals of the bestiary and include theories of information transfer and R.D. Guthrie's theories on the hunting prowess of young males.[8] Guthrie's theories and the climate hypothesis presented here are mutually compatible.

The great longevity of cave imagery over the millennia can be better understood when it is viewed as a cultural *meme*, and this idea is addressed below. From an evolutionary perspective, by analogy with Darwinian evolution, any cultural construct that can last 25,000 years, such as cave imagery, must have at least three characteristics:

- Longevity (of individual images)
- Ease of replication
- The ability to mutate or change

Such a long-lived, replicating, and mutating cultural unit has been called a *meme* by evolutionary biologist Richard Dawkins, similar to the French word même meaning the same.[11] Cave art is certainly repetitive, being dominated by the same species (bison and horse, for example). Each of these properties is addressed with regard to Upper Paleolithic cave imagery.

Longevity here refers to the lifespan of individual images—they must last long enough to be copied by others, otherwise they would disappear or have to be reinvented at a later time. For example, a drawing in the sand does not have a long enough lifespan to be reproduced by someone else at a later date. As noted already, Upper Paleolithic imagery in open-air rock shelters and the front of caves in daylight were probably much more common in the past but are now poorly represented, due to their destruction by the elements. Nevertheless, imagery that was engraved on hard rock surfaces could persist long enough to be replicated by subsequent generations. Assuming that each human generation undertook cave painting/engraving, the imagery would only have to last a generation in order for it to continue to be reproduced by the next generation. In the case of images that were painted, the paint recipes contained binders, such as feldspar and biotite, which prolonged their lifespan (see chapter 6). Engravings can be expected to last longer, depending on the hardness of the rock surface. Overall, because of the materials and techniques used, individual cave images can be expected to have a long lifespan, compared to other materials such as wood or animal hides.

The second criterion above is replication. The copying process should be reliable; otherwise the cultural construct (cave imagery in our case) will quickly mutate beyond recognition and lose any benefit it originally conferred on the culture. By analogy to biological organisms, it will die out or become extinct. A game the reader may have played when young helps illustrate replication. In this game, a person whispers a message to his or her neighbor sitting nearby, and the message is passed down the line to others. After four or five transmissions, the final message becomes totally garbled which can be amusing. This is referred to as the fidelity of transmission, and in this example, the fidelity is very poor. In biological evolution, the fidelity of replication of DNA is very good but not 100 percent, because of random mutations.

A slightly different game can be played, where the fidelity of transmission is much higher. In this game or experiment a person is given a paper origami model and asked to reproduce it, and the result is passed down the line.[12] As in the previous game, the end product is barely recognizable. A second experiment is performed, but this time instructions on how to fold the paper are also given. The subsequent paper models are much more successful in resembling the original.

Cave imagery, by its very nature, comes with its own set of instructions. The techniques of cave art are quite simple, involving mainly engraving, painting, drawing or shading combined in various ways (see chapter 6). Examination of any cave figure would reveal how it was constructed (engraved or painted or both) without the need of a teacher. Exceptions might be the very skillful three-dimensional clay modeling of bison at Le Tuc d'Audoubert cave (Ariège, France)[4] or the six giants of Henri Breuil, but such masterpieces are very rare. So cave imagery is more like the second origami experiment above than the first; it comes with its own set of instruction on how it can be replicated. This ensures good replication.

Leroi-Gourhan recognized different styles over time and style similarities within geographic regions, which means that replication does not have to be an exact copy. Most of the common large herbivore animals in cave imagery are recognizable, despite slightly different styles. Some direct dating by carbon-14 of animal figures thought to be contemporaneous based on style, yielded ages that differ by many thousands of years. These may be examples of faithful copies of earlier drawings where the original figure was imitated. Cave art fulfills the second criterion above—namely good fidelity in transmission or replication. If there is any doubt that cave art is not easily reproduced, the prehistorian Michel Lorblanchet, in an experiment, reproduced the large black frieze of mammoths, bison, and horses at Pech Merle in an hour.[13]

Transmission can be described as either vertical or horizontal. Vertical transfer refers to the parent-child situation or teacher-student relationship. In the case of vertical transmission of the imagery, children were no doubt involved in making cave imagery. Child footprints and hand stencils are preserved in several caves[2], and R. D. Guthrie devotes much discussion to the role of youth in cave art.[8] Horizontal transfer, on the other hand, is from one adult to another adult, as a member of a different group or tribe. If these groups were widely dispersed, information transfer between different groups may have been poor. Based on modern analogies with hunter-gatherer groups and excavation of Upper Paleolithic sites, campsites may have been composed of approximately 25–30 people.[14] Population density was thought to have been very low, in the range of one or two persons per 100 square km, except during the glacial maximum, when refuges (northern Spain and the Dordogne) became more densely populated.[14]

On a hypothetical square grid, these low population densities would place nearest neighbor base camps between 75 km (45 mi) and 60 km (37 mi) apart. At these distances, communication may have occurred only on a seasonal or longer time frame. Exotic raw materials (e.g., marine shells) at some Magdalenian sites do show that longer-distance procurement of raw materials occasionally occurred. It has been proposed that cave art sites, Altamira for example, were social gathering places where information and objects could be exchanged among such widely spaced groups.[15] This would have allowed horizontal replication of the cave imagery as an information source.[16]

The third characteristic of a meme is that it must have been able to undergo selective changes according to the conditions and the pressures exerted on the culture at any given time. If cave imagery provided no benefits to the entire culture, it would have died out earlier, just as biological organisms do not propagate harmful mutations for long periods. The main stress on the Upper Paleolithic hunter-gatherers over 25,000 years was climate change. From the Aurignacian to the end Magdalenian, they survived approximately twelve stadial and interstadial climate swings, each corresponding to winter temperature changes of about 15°C (27°F) to 20°C (36°F). As shown in chapter 8, the faunal assemblages depicted in caves accurately reflect these climate swings. This demonstrates the ability of the cave painters to evolve and change their depictions according to their environment.

These climate changes on such a short time scale (a few hundred years at most) would have had a profound effect on the lives of the Upper Paleolithic hunters over just four or five generations, assuming an adult hunter had a forty-year lifespan. New hunting techniques would have to be developed in the new landscape for the new fauna, which had different seasonal migrations, different behaviors, and different feeding and breeding patterns.

New hunting weapons would have to be developed and different stalking skills developed. If they did not adapt to the new conditions fast enough, they would become extinct, as did the Neanderthals before them. The only advantage these hunters had over the large mammals they hunted was cooperative hunting.[8] They relied on their intelligence and their creativity, experience, knowledge, and cunning. The advantage that cave art imagery conferred on these hunter groups was information—information to be shared with others in a cooperative effort to trap and kill their prey. Cave art fulfilled that role.

The meme analogy presented here does not imply a simple to complex progression through time, just as evolutionary biology does not. Extinctions have occurred in biology, and dead ends in cave art also likely occurred. The bizarre mammoths at La Baume Latrone[9] do not seem to have caught on elsewhere. The meme concept therefore is consistent with the interpretation of Paleolithic cave imagery in a post-Chauvet era of today, in which cave imagery cannot have followed Leroi-Gourhan's plan of a progression from simple to complex over time. Upper Paleolithic cave art satisfies the three conditions for a *meme*, namely longevity of individual images, ease of replication, and the ability to change.

The climate continued its unpredictable behavior during the Allerød-Bølling warming with the cold Dryas II in between, followed by the severe cold and dry climate of the Younger Dryas, which lasted 1,200 years. Finally, mild climate permanently returned 11,500 years ago, at the end of the Younger Dryas, and Upper Paleolithic cave art was not seen again. This particular cultural *meme* finally came to an end.

References cited

1. Bahn, P. G. and Vertut, J., 1997. *Journey through the Ice Age*. University of California Press.
2. Ucko, Peter, J. and Rosenfeld, A., 1967. *Paleolithic Cave Art*. World University Library.
3. Lawson, A. J., 2012. *Painted Caves: Paleolithic Rock Art in Western Europe*. Oxford University Press.
4. Clottes, J. and Lewis-Williams, D., 1998. *The shamans of Prehistory*. Abrams, New York.
5. Bahn, P. G., 1995. "Cave art without the caves." *Antiquity*, 69, 231–237.

6. Halverston, J., 1987. "Art for Art's sake in the Paleolithic." *Current Anthropology*, 28, 63–89.
7. Reinach, S., 1903. "L'art et la Magie. A propos des peintures et des gravures de l'age du Renne." *L'Anthropologie*, 14.
8. Guthrie, R. D., 2005. *The Nature of Paleolithic Art*. University of Chicago Press.
9. Leroi-Gourhan, A., 1967. *Treasures of Prehistoric Art*. Abrams, New York.
10. Lewis-Williams, D. and Dowson, T.A., 1988. "The signs of all times: Entoptic phenomena in Upper Paleolithic art." *Current Anthropology*, 29, 201–245.
11. Dawkins, R., 1989. *The Selfish Gene*. Oxford University Press.
12. Atran, S., 2001. "The trouble with memes: Inference versus imitation in cultural creation." *Human Nature*, 12, 351–381.
13. Lorblanchet, M., 1980. "Peindre sur les parois des grottes." *Dossier de l'Archéologie*, 46, 33–39 quoted in Conkey, M. W., 1987. "New Approaches in the search of meaning? A review of research of Paleolithic art." *J. Field Archaeology*, 14, 413–430.
14. Bocquet-Appel, J-P., 2000. "Population kinetics in the Upper Paleolithic in Western Europe." *J. Archaeological Science*, 27, 551–570.
15. Conkey, M. W., 1980. "Identification of prehistoric hunter-gatherer aggregation sites: the case of Altamira." *Current Anthropology*, 21, 609–630.
16. Mithen, S. J., 1988. *Looking and learning: Upper Paleolithic art and information gathering*, 19, 297–327.

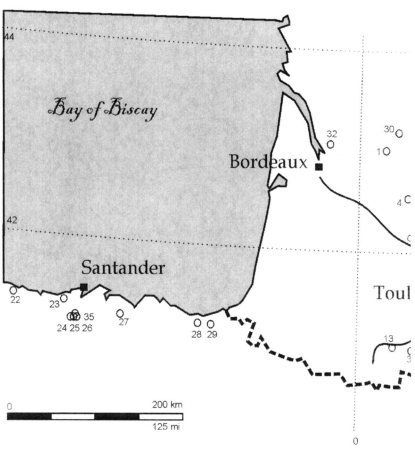

Cave locations mentioned in the text

1. Gabillou
2. Rouffignac
3. Lascaux
4. Combarelles
5. Font-de-Gaume
6. Bernifal
7. Cougnac
8. Pech-Merle
9. Pergouset
10. Chauvet
11. Chabot
12. Ebbou
13. Gargas
14. Marsoulas
15. Les Trois-Frères
16. Le Tuc d'Audoubert
17. Le Portel
18. Niaux
19. La Vache
20. La Baume Latrone
21. Cosquer
22. Tito Bustillo
23. Altamira
24. La Pasiega
25. Las Monedas
26. Las Chimeneas
27. Covalanas
28. Ekain
29. Altxerri
30. Teyjat
31. Villars
32. Pair-non-pair
33. Mas D'azil
34. Montespan
35. El Castillo

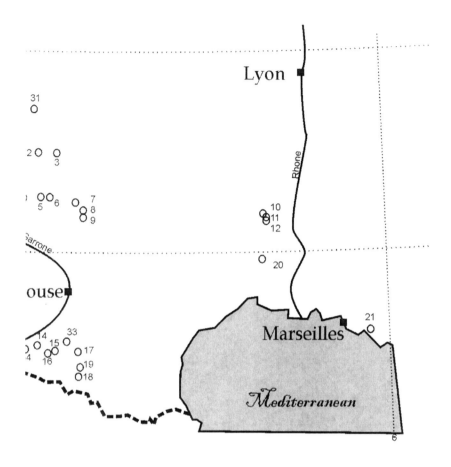

Index

A

African elephants 103
Agassiz Louis 1
Allerød-Bølling
    interstadials 27, 28, 40, 92, 111
Altamira
    bison 73
    carbon-14 72
    climate curve 93
    stylistic dating 72
Aquitaine 45
Aquitane 100
art for art's sake 5, 105
Aujoulat N. 64
Aurignacian
    Abri Pataud 30
    calibrated years 9
    Dordogne 30
    stone tools 15
aurochs
    wild cattle 43

B

Baume Latrone 40, 61, 102
Baume Latrone
    bizzare mammoths 111
Bering strait 32

bison 42
    DNA 42
    Lascaux 42
bones and cave art
    French/Spanish caves 83
Breuil H 13, 44, 66, 67, 71, 72, 93, 106, 109
Buckland W 16

C

carbon-14
    AMS 70
    calibration 69
    confidence interval 69
    contamination 70
    error bars 79
    half life 68
    Libby WF 65
Castillo El 12, 41, 54, 75, 89, 91
    hand stencils 12
cave art
    as meme 108
    French and Spanish 50
Chabot
    discovered 1
Chauvet
    age 9, 13
    charcoal 71

Chauvet-type signs 99
climate curve 94
cold faunal assemblage 87
cold period 55
discovered 57
techniques 64
Child footprints 109
Chimeneas
    Lascaux interstadial 89
    temperate assemblage 88
CLIMAP 25
climate index variable 88
Clottes J 2, 9, 12, 57, 61, 78
Clovis 17
Côa Valley
    petroglyphs 68
Cosquer
    climate curve 94
    phase I 89
    phase II 95
Cougnac
    discovered 91
    paint recipes 63

D

Dawkins R
    cultural meme 108
dust levels 32. *See* GISP2

E

Ebbou 102
Echegaray J 90, 91
Ekain
    cave art 78
    kitchen bones 78

F

factor analysis 49
Four Hundred Centuries of Cave Art
    13, 66. *See* Breuil H
French
    structuralism 107

G

giant deer
    Irish elk 44
GISP2 27, 29, 30, 31, 32, 33, 88, 90, 95
    dust levels 32
    methane 30
Gönnersdorf 40, 77, 78, 83
    kitchen refuse 77
    mammoths 77
Gravettian
    burials 17
    calibrated years 9
    Cougnac 92
    Pech Merle 93
    red lady 16
greenhouse gases
    Arrhenius S 26
    carbon dioxide and methane 4
Greenland
    ice core. *See* GISP2
Guthrie R D 6, 39, 40, 109
    male hunting 107

H

Hand stencils 64
Heinrich event 29, 69
Herzog W
    Chauvet film 57
Holocene 4, 19, 35, 38, 39, 44, 56, 68, 83, 105
    Côa Valley 68
    Levantine art 83
Homo sapiens 3, 11, 14
horse
    cold-adapted 54
    Equus caballus 43
    przewalski 43
hunting magic hypothesis 106

I

ibex
    Capra ibex 44
    Massif Central 87
    Pyrenees 87
    rocky slopes 54
ice core
    Antarctica 3
    carbon dioxide 27
    glacials/interglacials 28
    Greenland 3, 26
    interstadials 29
    methane 27
    Milankovitch cycles 27
    National laboratory, Denever 23
    oxygen isotopes 28
    timescale 24
    Vostock 26
    Younger Dryas 28
interstadials 28

K

Kesslerloch 1, 78, 85
    excavation 78

L

Lascaux
    Abbe Breuil 66
    Aujoulat N 64
    crayons 63
    discovery 2
    excavation 2
    reindeer bones 77
    temperate assemblage 87
    twisted perspective 66
    warm period 55
Lascaux interstadial 88
Last Glacial Maximum
    beginning 28
    extent of ice 32
    frozen caves 69

herbivores 45
length 58
mammoth migrations 54
methane levels 4
population 101
temperature 4
tundra-steppe 38
Laugerie interstadial 88
La Vache
    arctic fox 37
    kitchen refuse 77
    paint recipes 64
Leroi-Gourhan 13, 50, 51, 52, 61, 66, 67, 85, 90, 91, 92, 93, 94, 99, 106, 109, 111
    styles 66
Le Tuc d'Audoubert 63, 109
Libby WF
    carbon-14 65
    Nobel Prize 65
loess 32. *See* dust levels
Lorblanchet M
    Cougnac 63

M

Magdalenian
    Altamira 73
    calibrated years 9
    cave art % 17
    Ekain 80
    hunting camps 17
mammoth caves 100
mammoth migrations 101
Mammoth Steppe
    tundra-steppe 38
meme
    Dawkins R 108
methane
    atmosphere 30
    GISP2 31
    wetlands 30

Milankovitch M
   cycles 26
Monedas, Las 89

N

Neanderthals
   clothes 14
   extinction 11
Neolithic 3, 4, 14
Niaux
   discovered 93
   paint recipes 64
   Salon Noir 93
Niaux and La Vache 80
North Atlantic
   Gulf Stream 29

O

Ötzi the ice man
   clothes 14

P

paint recipes
   Cougnac 64
   La Vache 64
   Niaux 64
Pasiega La 91
Pech Merle 93
   spotted horses 93, 107
Pincevent 18, 19, 44
Poisson statistics 96
polar desert 101
pollen
   Chauvet 57
   grasses 57
   Lascaux 57
   Mediterranean 33
   trees 57
portable art
   definition 1
Portel, Le 91
principle component analysis 49

R

red deer 43
Reinach S 6
reindeer 44
Rouffignac
   discovered 2
   engravings 63
   faunal assemblage 99

S

Sautuola M
   Altamira 72
Shamanism 107
Solutrean
   bone needles 15
   calibrated years 9
   Chabot 17
   ibex hunting 55
   stone tools 17
Spain
   mammoths 101
   refuge 101
stadials 28
statistical tests 59
statistics
   Poisson 96
styles
   Abbe Breuil 66
   Leroi-gourhan 66
stylistic dating
   Altamira 71
   Chauvet 67

T

theories of cave art 5, 105
tundra-steppe
   temperatures 38
   wooly mammoth 38
twisted perspective
   Lascaux 66

Tyndall J
    greenhouse gases 26

V

Verberie 18, 19, 79
von Humboldt A
    German naturalist 37
Vostock
    ice core 26, 28

W

wooly mammoth
    diet 39
    Dima 39
    global range 41, 99
    Mammuthus primigenius 39
    migration 101
wooly rhinoceros 42

Y

Younger Dryas
    end of 28, 111
    length 19
    refugia 19

About the Author

Kieran D. O'Hara, a native of Dublin, Ireland. He spent thirty years at the University of Kentucky as a faculty member in the Department of Earth and Environmental Sciences, where he is a professor emeritus. He lives in Lexington, Kentucky.

CPSIA information can be obtained at www.ICGtesting.com
Printed in the USA
LVOW10s1759270915

455916LV00027B/1248/P